Agent Based Approach For Supply Chain Management

Dr. M S Uppin

INDIA • SINGAPORE • MALAYSIA

Notion Press

Old No. 38, New No. 6
McNichols Road, Chetpet
Chennai - 600 031

First Published by Notion Press 2018
Copyright © M S Uppin 2018
All Rights Reserved.

ISBN 978-1-64249-483-9

This book has been published with all reasonable efforts taken to make the material error-free after the consent of the author. No part of this book shall be used, reproduced in any manner whatsoever without written permission from the author, except in the case of brief quotations embodied in critical articles and reviews.

The Author of this book is solely responsible and liable for its content including but not limited to the views, representations, descriptions, statements, information, opinions and references ["Content"]. The Content of this book shall not constitute or be construed or deemed to reflect the opinion or expression of the Publisher or Editor. Neither the Publisher nor Editor endorse or approve the Content of this book or guarantee the reliability, accuracy or completeness of the Content published herein and do not make any representations or warranties of any kind, express or implied, including but not limited to the implied warranties of merchantability, fitness for a particular purpose. The Publisher and Editor shall not be liable whatsoever for any errors, omissions, whether such errors or omissions result from negligence, accident, or any other cause or claims for loss or damages of any kind, including without limitation, indirect or consequential loss or damage arising out of use, inability to use, or about the reliability, accuracy or sufficiency of the information contained in this book.

Contents

List of Figures ix

List of Tables xi

List of Abbreviations xiii

1. **Introduction** 1
    1.1   Background 1
    1.2   Overview of the Work Carried Out 5
        1.2.1   Customer Requirement Module (CRM) 8
        1.2.2   Market Survey Module (MSM) 8
        1.2.3   Product Design and Development Module (PDDM) 8
        1.2.4   Material Resource Planning Module (MRPM) 8
        1.2.5   Purchasing Activity Module (PAM) 9
        1.2.6   Process Planning Module (PPM) 9
        1.2.7   Production Scheduling Module (PSM) 9
        1.2.8   Assembly and Testing Module (ATM) 9
        1.2.9   Quality Control Module (QCM) 10
        1.2.10 Customer Care and Sales Module (CSM) 10
        1.2.11 Financial Activities Module (FAM) 10
        1.2.12 Storage and Material supply module (SMM) 10
        1.2.13 Subcontracting Module (SUCM) 11
        1.2.14 Transportation and Distribution Module (TDM) 11

	1.2.15	Distributors and Retailers module (DRM)	11
	1.2.16	Supporting Databases	11
1.3		Outline of the Theses	12
2.	**Literature Review**		**15**
2.1		History of SCM	17
2.2		Supply Chain	18
2.3		Supply Chain Management	22
2.4		Information Technology and SCM	27
2.5		SCM and Internet	32
2.6		SCM and E-marketplace	37
2.7		SCM And Mutual Trust with Partners	39
2.8		Agent Technology	41
2.9		SCM and Information sharing	46
2.10		Multi Agent Systems	50
2.11		Significance of Purchasing in SCM	55
2.12		Significance Issues on Information sharing	57
2.13		The Problem Formulation	60
3.	**Concepts of Agent Based Approach.**		**61**
3.1		Description of Agent	61
3.2		Properties of an Agent	64
3.3		Generic Agent	66
3.4		Significance of Software Agents	69
3.5		Agents as Design Metaphor	71
3.6		Agents as Source of Technology	71
3.7		Agents for Simulation.	72
3.8		Trust and Reputation	72
3.9		Interaction Level Coordination	74
3.10		Negotiation	74
3.11		Communication	76
3.12		Agent Based Manufacturing	77

	3.13	Applications of Agent Technology in the Manufacturing Domains	79
		3.13.1 Engineering Design	79
		3.13.2 Process Planning	80
		3.13.3 Production Planning and Resource Allocation	81
		3.13.4 Production Scheduling and Control	81
		3.13.5 Process Control, Monitoring and Diagnosis	82
		3.13.6 Enterprise Organization and Integration	83
		3.13.7 Production in Networks	84
		3.13.8 Assembly and Life-Cycle Management	84
	3.14	Benefits of Agent Based Approach for Manufacturing	85
4.	**Agents for MRP, Process Planning and Scheduling**		**87**
	4.1	Fundamentals of MRP for Recognition of Necessary Agents	87
		4.1.1 Inputs to MRP	89
		4.1.1.1 Master Production Schedule	89
		4.1.1.2 Bill of Material	90
		4.1.1.3 Inventory Data	90
		4.1.2 Conditions for Implementation	91
	4.2	Agent Based MRP	91
		4.2.1 Storage Agent	91
		4.2.2 Agent for the Generation of Master Production Schedule (MPSA)	92
		4.2.3 Agent for Rough Cut Capacity Planning (RCPA)	92
		4.2.4 Product Data Agent (PDA)	92
		4.2.5 Agent for Bill of Material (BOMA)	94
		4.2.6 Agent for Determination of Inventory Status files (ISA)	94
		4.2.7 Agent for Time Phasing (TPA)	94
	4.3	Fundamentals of Process Planning	98
		4.3.1 Steps in Process Planning	98
		4.3.2 Basic Information Required for CAPP	99
	4.4	Agent Based Process Planning	101
		4.4.1 Data and Knowledge base Management Agent (DKMA)	103
		4.4.1.1 Product Database (PD)	103

		4.4.1.2	Coded Information (CI)	103
		4.4.1.3	Operational Database (OD)	103
		4.4.1.4	Resource Database (RD)	103
		4.4.1.5	Machining Database (MCD)	104
		4.4.1.6	Machining Knowledgebase (MKB)	104
	4.4.2	CAD Data Extraction Agent (CDEA)		105
	4.4.3	Feature Recognition and Presentation Agent (FRPA)		105
	4.4.4	Operation Selection Agent (OPSA)		106
	4.4.5	Machine Tool, Cutting Tool and Fixture Selection Agent (MCFA)		107
	4.4.6	Process Parameter Selection Agent (PPSA)		108
	4.4.7	Setup Plan Generation Agent (SPGA)		109
	4.4.8	Operations Sequencing Agent (OSQA)		109
	4.4.9	NC Program Generating Agent		109
	4.4.10	Report Generation and Presentation Agent (RGPA)		110
4.5	Scheduling			110
	4.5.1	Scheduling Algorithm		111
	4.5.2	Capacity Requirement Planning		113
4.6	Agents for Scheduling			113
	4.6.1	Scheduling Agent (SCHA)		113
	4.6.2	Available Resource Information Agent (ARIA)		114
	4.6.3	Capacity Requirement Planning Agent (CRPA)		114
	4.6.4	Sequence Verification Agent (SQVA)		115
	4.6.5	Standard Time Calculation Agent (STCA)		115
	4.6.6	Schedule Execution Agent (SEXA)		115
	4.6.7	Shop Floor Control Agent (SFCA)		115
5.	**Multi Agent Framework for Purchasing Process**			**117**
5.1	Steps in Purchase Process			117
	5.1.1	Requisition Generation & Approval		118
	5.1.2	Identification of Suppliers		119
	5.1.3	Evaluation of Suppliers & Bidding/Negotiating		120
	5.1.4	Selection of Suppliers		123

	5.1.5	Purchasing Approval	124
	5.1.6	Purchase Order Release	124
	5.1.7	Expediting and Delivery from Supplier	125
	5.1.8	Supplier Invoice Paid	126
	5.1.9	Update Supplier Information	127
5.2	Proposed Agents for the Automation of Purchasing Process		129
	5.2.1	AGENT 1 – Availability Checking & Material Requisition Agent (ACMRA)	129
	5.2.2	AGENT 2 – List of Items to Purchase Agent (LIPA)	132
	5.2.3	AGENT 3 – Quotation Sending Agent (QSA)	133
	5.2.4	AGENT 4 – Quotation Receiving and listing Agent (QRA)	134
	5.2.5	AGENT 5 – Vendor Rating Agent (VRA)	134
	5.2.6	AGENT 6 – Purchase Order Generation Agent (POGA)	136
	5.2.7	AGENT 7 – Purchase Order Tracking Agent (POTA)	137
	5.2.8	AGENT 8 – Material Receipt & Stock Updating Agent (MRSUA)	137
	5.2.9	AGENT 9 – Vendor data Updating Agent (VUA)	138
	5.2.10	AGENT 10 – Invoice Clearance Agent (ICA)	138

6. **Execution of Agents for Performing Activities Related to Purchasing** .. 145

7. **Results and Discussions** 167

8. **Conclusions** 189
   - 8.1 Summary 189
   - 8.2 Salient Features 190
   - 8.3 Limitation of the Present Work 192
   - 8.4 Scope for Future Work 192

References *195*

Author's Publications *207*

LIST OF FIGURES

Figure No.	Figure Name	Page No.
Fig. 1.1:	Modular Structure of Agent Based Model for Supply Chain Functions	7
Fig. 2.1:	Supply Chain Management-An Overview	24
Fig. 2.2:	Typical Information Technology for SCM	29
Fig. 2.3:	Supply Chain and Information Coupling	30
Fig. 2.4:	Typical Agent	42
Fig. 2.5:	Typical Supply Chain Network and Agents	49
Fig. 4.1:	Schematic View of Material Requirement Planning	88
Fig. 4.2:	Constituents for Master Production Schedule	90
Fig. 4.3:	Agents for MRP Activities	93
Fig. 4.4:	Logic for Time Phasing	95
Fig. 4.5:	Agents Identified for the Generation of Process Plan	102
Fig. 4.6:	Scheduling Agents	116
Fig. 5.1:	Purchasing Process	128
Fig. 5.2:	Agents Identified for the Generation of Purchasing Process	130
Fig. 5.3:	Flow Chart Describing Logic of Executing Agents for Purchasing	139
Fig. 6.1:	Log-In Screen	146
Fig. 6.2:	Welcome Form for Storage Database	147

Fig. 6.3:	Welcome Form for Checking Available Material	148
Fig. 6.4:	Form for Checking Availability of a Specific Material	148
Fig. 6.5:	Form Displaying the Details of Specific Material Form Storage Database	149
Fig. 6.6:	Form for Checking All Available Material List	150
Fig. 6.7:	Form for Submitting Request for Supply of Material from Store	151
Fig. 6.8:	Material Requisition Form	152
Fig. 6.9:	Form for Entering Vendor Specific Data by Authorized User	154
Fig. 6.10:	Form for Entering Material Specific Data by the Authorized User	154
Fig. 6.11:	Form Displaying List of Material to Be Purchased	155
Fig. 6.12:	User Screen for Vendor Database	156
Fig. 6.13:	Screen for Material with Vendors	157
Fig. 6.14:	Requisition Form for a Quotation	158
Fig. 6.15:	Quotation Form	159
Fig. 6.16:	Form Displaying Vendor Rating Data for a Specific Material	161
Fig. 6.17:	Screen for Material with Vendors	161
Fig. 6.18:	Purchase Order	162
Fig. 6.19:	Form for Expatiation	163
Fig. 6.20:	Form for Acceptance or Rejection of Materials Received	164
Fig. 6.21:	Form for Material Received	165
Fig. 7.1:	BOM for Example Part	169
Fig. 7.2:	Form Displaying List of Material to Be Purchased	176
Fig. 7.3:	Screen for Vendor Database	177
Fig. 7.4:	Form for Material with Vendors	180
Fig. 7.5:	Form for Inviting a Quotation	181
Fig. 7.6:	Sample Screen for Compiled Vendor Rating for a Material	183
Fig. 7.7:	Sample Form for Material List with Selected Vendors	187
Fig. 7.8:	Purchase Order for RM1	188

LIST OF TABLES

Table No.	Table Details	Page No.
Table 5.1:	Contents of Item Database	131
Table 5.2:	Determination of Quantity and Due Date for Item Required for Multiple Departments	132
Table 5.3:	Contents of Vendor Database	133
Table 5.4:	Vendor Performance Rating Calculation	136
Table 7.1:	MSP for Example Part	168
Table 7.2:	Time Phased Requirement of Different Components of Example Part	168
Table 7.2.1:	Planned Requirement for Chair (C)	168
Table 7.2.2:	Planned Requirement for Seat Assembly (SA)	170
Table 7.2.3:	Planned Requirement for Hands Set (HS)	170
Table 7.2.4:	Planned Requirement for Legs Assembly (LA)	170
Table 7.2.5:	Planned Requirement for Seats Plane (SP)	170
Table 7.2.6:	Planned Requirement for Fixture (FX1)	171
Table 7.2.7:	Planned Requirement for Hands (HA)	171
Table 7.2.8:	Planned Requirement for Fixture Set 2 (FX2)	171
Table 7.2.9:	Planned Requirement for Legs (LG)	171

Table 7.2.10:	Planned Requirement for Rod (RD)	172
Table 7.2.11:	Planned Requirement for Cover (CV)	172
Table 7.2.12:	Planned Requirement for RM1	172
Table 7.2.13:	Planned Requirement for RM2	172
Table 7.2.14:	Planned Requirement for RM3	173
Table 7.3:	List Items to Be Manufactured	173
Table 7.4:	List of Items to Be Purchased	174
Table 7.5:	List and Address of Vendors	177
Table 7.6:	List of Vendors for RM1	182
Table 7.7:	Vendor Rating for RM1	182
Table 7.8:	Vendors for RM2	183
Table 7.9:	Vendor Rating for RM2	184
Table 7.10:	Vendors for RM3	184
Table 7.11:	Vendor Rating for RM3	184
Table 7.12:	Vendors for Seating Plane (SP)	185
Table 7.13:	Vendor Rating for Seating Plane (SP)	185
Table 7.14:	Vendors for Fixture (FX)	185
Table 7.15:	Vendor Rating for Fixture (FX)	186
Table 7.16:	Vendors for Rod (RD)	186
Table 7.17:	Vendor Rating for Rod (RD)	186

LIST OF ABBREVIATIONS

Abbreviations	Description
ABMS	Agent-based modeling and simulation
ACMRA	Availability Checking & Material requisition Agent
AI	Artificial Intelligence
AIM	Assembly & Inspection Module
AMT	Advanced Manufacturing Technology
AOP	Agent-Oriented Programming
AQ	Average Quantity
ARIA	Available resource Information Agent
ASL	Approved Supplier List
ATM	Assembly and Testing Module
B2B	Business to Business
B2C	Businesses to Consumers
BDI	Belief-Desire-Intention
BOM	Bill of Material
BOMA	Bill of Material Agent
BPR	Business Process Re-engineering
CA	Capacity Analysis
CAD	Computer Aided Design

Contd...

Abbreviations	Description
CAE	Computer Aided Engineering
CAM	Computer Aided Manufacturing
CAPP	Computer Aided Process Planning
CDEA	CAD Data Extraction Agent
CEO	Chief Executive Officer
CI	Coded Information
CIM	Computer Integrated Manufacturing
COA	Customer Order Agent
CPFR	Collaborative Planning Forecasting and Replenishment
CRM	Customer Requirement Module
CRM	Customer requirement Module
CRP	Capacity Requirement Planning Agent
CRPA	Capacity Requirement Planning Agent
CSM	Customer Care and Sales Module
DAI	Distributed Artificial Intelligence
DC	Distributed Computing
DF	Directory Facilitator
DKMA	Data and Knowledge base Management Agent
DRM	Distributor & Retailer Module
DUDA	Data Updating Display Agent
E	Electronic
EDI	Electronic Data Interchange
EM	Electronic Market
ERP	Enterprise Resource Planning
E-SCM	Electronic Supply Chain Management
FAM	Finance Activity Module
FIPA	Foundation for Intelligent Physical Agents
FLEX_FI	Flexibility in delivery quantity and due date with full information – sharing

Abbreviations	Description
FLEX_NI	Flexibility in delivery quantity and due date without information – sharing
FLEX_PI	Flexibility in delivery quantity and due date with partial information
FMS	Flexible Manufacturing System
FRPA	Feature Recognition and Presentation Agent
GA	Genetic Algorithm
ICA	Invoice clearance Agent
IDE	Integrated Development Environment
IS	Information System
ISA	Inventory status files Agent
IT	Information Technology
JIT	Just In Time
KAoS	Knowledge-able Agent-oriented System
KQML	Knowledge Query and Manipulation Language
LIM	List of Items for Manufacturing
LIP	List of Items for Purchasing
LIPA	List of Items to Purchase Agent
LIS	List of Items for Subcontracting
MAS	Multi Agent Systems
MCD	Machining Database
MCFA	Machine Tool Cutting Tool and Fixture Selection Agent
MES	Manufacturing Execution Systems
MF	Market forecast
MKB	Machining Knowledgebase
MPSA	Master Production Schedule Agent
MPSM	Master Production Scheduling Module
MRF	Material Requisition Form
MRP	Material Requirement Planning

Contd...

Abbreviations	Description
MRPM	Material Resource Planning Module
MRSUA	Material Receipt & Stock Updating Agent
MSM	Market Survey Module
NC	Numerical control
NCPA	NC Program Generating Agent
OD	Operational Database
OMG	Object Management Group
OMM	Overall Maintenance Module
OOP	Object-Oriented Programming
OPSA	Operation Selection Agent
OSQA	Operations Sequencing Agent
PAM	Purchasing Activity Module
PD	Product Data base
PDA	Personal Digital Assistants
PDDM	Product Design and Development Module
PDDM	Product design & Development Module
PO	Purchase Order
POGA	Purchase Order Generation Agent
POTA	Purchase Order Tracking Agent
PPM	Process Planning Module
PPSA	Process Parameter Selection Agent
PSM	Production Scheduling Module
QCM	Quality Control Module
QR	Quick Response
QRA	Quotation Receiving & listing Agent
QSA	Quotation Sending Agent
RCPA	Rough cut Capacity Planning Agent
RD	Resource Database

Abbreviations	Description
RGPA	Report Generation and Presentation Agent
SCADAS	Supply Chain Agent Decision Aid System
SCHA	Scheduling Agent
SCI	Supply Chain Integration
SCM	Supply Chain Management
SDM	Storage Data Module
SDM	Storage Database Module
SEXA	Schedule Execution Agent
SFA	Sales Forecasting Agent
SFCA	Shop Floor Control Agent sharing
SPGA	Setup plan Generation Agent
SQC	Statistical Quality Control
SQVA	Sequence Verification Agent
STCA	Standard Time Calculation Agent
STO	Stochastic Model
SUCM	Subcontracting Activity Module
SUCM	Subcontracting Module
SVL	Selected Vendor List
TAC SCM	Trading Agent Competition Supply Chain Management
TDM	Transportation & Distribution Module
TPA	Time Phasing Agent
TQM	Total Quality Management
VMI	Vendor Managed Inventory
VRA	Vendor Rating Agent
VUA	Vendor data Updating Agent
WWW	World Wide Web

CHAPTER 1

INTRODUCTION

1.1 BACKGROUND

Today's world market environment is rushing towards its total globalization, and is characterized by ever increasing pace in production and decreasing product cycle times. The common trends that are observed in today's manufacturing systems are: numerous competitors, market globalization, a steadily increasing complexity of business processes, a highly turbulent production environment, development and application of new product technologies, increase in product variety and decrease in product volumes, increase in delivery reliability and products are expected to be of low-cost, high quality and high reliability. Therefore businesses that are more responsive to market changes and more sensitive to customers needs are more likely to survive and thrive in such environment.

In order to support its global competitiveness and rapid market responsiveness, an individual manufacturing enterprise has to be integrated not only with its related systems such as purchasing, design, production, planning and scheduling, control, transport, resources, personnel, materials, quality, etc., but also with its partners, suppliers and customers through heterogeneous software and hardware environments[1]. Supply Chain encompasses all those activities needed to design, make, deliver and use a product or service[2,3]. Hence, the pace of change and the uncertainty about how markets will evolve has made it increasingly important for companies to be aware of the supply chains they participate in and to understand the roles that they play. Collaboration among entities within the supply chain has a great impact on the system performance[4].

Those companies that learn how to build and participate in strong supply chains will have a substantial competitive advantage in their markets. With customers becoming more demanding in their requirement of services from the suppliers, the construction of an efficient and integrated supply-chain has assumed paramount importance[5].

A supply chain typically extends across the multiple enterprises including suppliers, manufacturers, transportation carriers, ware houses, retailers as well as customers and entails sharing forecast, order, inventory, and production information to better coordinate management decisions at multiple points throughout the extended enterprise[5]. Most importantly they are embodied with information system necessary to monitor all these activities. Since companies seek to provide products and services to customers faster, cheaper and with better quality, the supply chain system is becoming recognized as strategically critical aspect of the firm.

Partnering programs, such as vendor managed inventory (VMI), collaborative planning, forecasting, and replenishment (CPFR), quick response (QR), and continuous replenishment, increases the level of interaction between companies. All of these programs share a common element, sharing information and decreasing the uncertainty built into information that flows back up the supply chain[6].

To optimize performance, supply-chain functions must operate in a coordinated manner[7]. But the dynamics of the enterprise and the market make this difficult: Materials do not arrive on time, production facilities fail, and workers are ill, customers change or cancel orders, and so forth, causing deviations from the plan. In some cases, these events may be dealt with locally; that is, they lie within the scope of a single supply-chain function. In other cases, the problem cannot be locally contained and modifications across many functions are required. Consequently, the supply-chain management (SCM) system must coordinate the revision of plans or schedules across supply-chain functions. The timely flow of information and accurate coordination of decisions ultimately determines the success of supply chain.

Several studies have shown that there is a positive effect of information sharing along the supply chain and better integration as well as coordination among the different nodes in the supply chain through use of information technology has also helped improve performance in these systems. In the ideal case, a supply chain facilitates the availability of the right amount of the right product at the right price at the right place with minimal inventory across the network.

Over the past two decades a large number of manufacturing systems, production models and philosophies have been developed by exploring the advanced manufacturing techniques that satisfy the requirements of global competition and changing needs of the customer. Typical examples are CAD/CAM/CAE/CAPP/CIM, FMS, and Design for manufacturing, Agile manufacturing, Intelligent manufacturing systems, Agent based manufacturing system.

The common characteristics of all these manufacturing systems/ models are:

a) Application of computers to all most all the activities right from design, planning, manufacturing, quality control to customer services.

b) Automation of most of the activities.

c) Integration of different activities.

d) Application of large amount of data stored in different databases.

e) Information Technology for the successful implementation of these manufacturing systems and production strategies[8].

Therefore it can be concluded that, Information Technology (IT) plays a major role in the formation of the supply chain also in enhancing the effectiveness and efficiency of supply chain activities. The decision support provided by IT products are expected to help the decision-makers in the development of the supply chain process and in implementation. This is due to the fact that, the supply chain usually includes different units of a company and also more than one company and the communication and information sharing between units/companies at the supplier-customer interface is critically important for overall supply chain performance[9,10,11]. From information technology perspective SCM synchronizes a set of interrelated activities across the multiple organization boundaries with different computing platforms.

However, the major problem facing manufacturing organizations is how to provide efficient and cost-effective information sharing systems due to the complexity of supply chain networks coupled with the complexity of individual manufacturing systems within supply chains. Although information plays a major role in effective functioning of supply chain networks considerable work has not been carried out regarding the dynamics of supply chains and how data collected in these systems can be used to improve their performance

Hence, there is a need to identify software development process which can support the management of heterogeneous or diversified information used for modeling supply chain components. In this regard Agent technology is found to be most promising in satisfying the above needs. An agent is a software entity that has a set of protocols which govern the operations of the manufacturing entity, a knowledge base, an inference mechanism and an explicit model of the problem to solve. It is autonomous, goal-oriented software that operates asynchronously, communicating and coordinating with other agents as needed.

Agent technology provides a natural way to address the issues related to information sharing in order to design and implement distributed intelligent manufacturing environments and provides software architecture for managing the supply chain. It views the supply chain as composed of a set of intelligent agents, each responsible for one or more activities in the supply chain and each interacting with other agents in planning and executing their responsibilities.

A single business unit in a supply chain is a company, which operates autonomously to achieve its own objectives, organizing its possible actions into plans on the basis of the available knowledge. A supply chain can be considered as a network of autonomous business units aiming at the procurement, manufacturing and distribution of related products.[12]

Keeping in mind the above-mentioned analogies between a company in a supply chain and an agent, it becomes clearer that the Multi-Agent System paradigm is a valid approach for modeling supply chain networks and for implementing supply chain management applications. The benefits of adopting agent technology in supply chain management are several; especially it is possible to satisfy all the issues identified in the proposed work regarding efficient flow of information in a supply chain.

1.2 OVERVIEW OF THE WORK CARRIED OUT

With this idea the present work has been carried out is concerned with the application of basic concepts of Agent based technology for the flow of information between various components of the supply chain formed for a manufacturing organization with following objectives;

1. To identify the issues related to information sharing as the most critical factor of supply chain activities.
2. To evaluate the multi agent system concepts as effective means of addressing the problems related to information sharing between various units of supply chain with in same organization as well as between different organizations.
3. To formulate a agent based frame work for the supply chain management activities leading to effective sharing of information.
4. To develop a multi agent system to perform certain major activities of supply chain formed for a manufacturing organization and demonstrate the sharing of information, in such a manner that it can prove an initial and major step towards enhancing the effectiveness of supply chain activities.
5. Providing significant contribution to modern manufacturing enterprises to face global competitiveness.

The work includes development of different Agents in the form of autonomous software, each capable of communicating with other Agents to share the information. Each pair of Agents communicates with each other by sending well defined messages and also responding suitably to any message received.

For the purpose of effective information sharing agents are grouped into different modules, each group of agents in a module perform a major function of the enterprise. The list and description of all the proposed modules are given below and their integrated activities in Figure. 1.1.

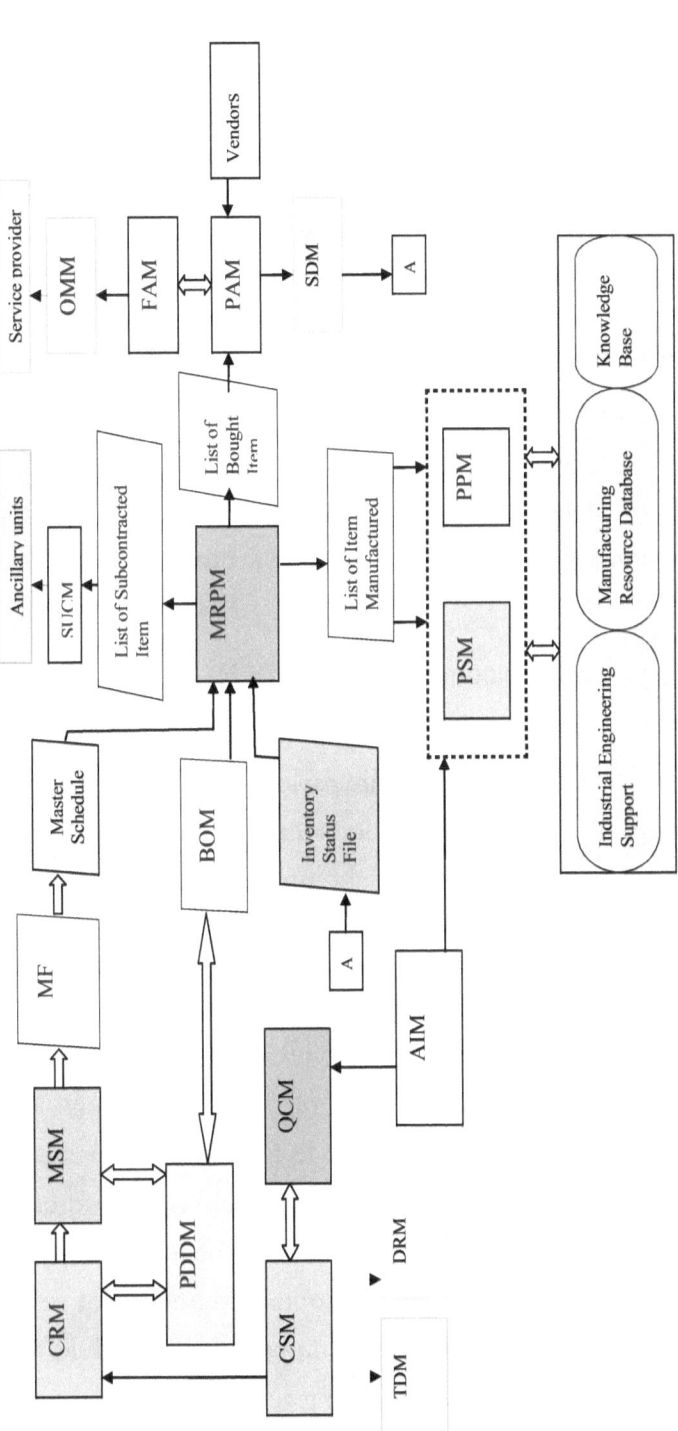

Fig. 1.1: Modular Structure of Agent Based Model for Supply Chain Functions

CRM- Customer Requirement Module, MSM-Market Survey Module, MF- Market forecast, PDDM- Product Design And Development Module, BOM-Bill Of Material, MRPM-Material Resource Planning Module, FAM-Finance Activity Module, PAM-Purchasing Activity Module, CSM- Customer Care and Sales Module, QCM-Quality Control Module, AIM- Assembly & Inspection Module, PSM- Production Scheduling Module, PPM- Process Planning Module, SDM- Storage Database Module, OMM-Overall Maintenance Module, TDM-Transportation & Distribution Module, DRM-Distributor & Retailer Module, SUCM- Subcontracting Activity Module.

1.2.1 Customer Requirement Module (CRM)

This module is designed to perform the tasks of gathering the information regarding demands from the regular as well as new customers. The module is efficient in compiling both qualitative and quantities information related to customer requirements.

1.2.2 Market Survey Module (MSM)

This module is concerned with activities related to Market and customers in terms of demands and requirements. It is responsible for analyzing the data from market as well as customer feedbacks obtained form CRM and convert this information into production Plan for the near future.

1.2.3 Product Design and Development Module (PDDM)

This module is concerned with design and development of products with the help of Computer Aided Design software packages along with the data acquisition from customer point of view. Some of its Agents collects and stores all the required information about the product for any reference before, during, and after manufacturing activities.

1.2.4 Material Resource Planning Module (MRPM)

This module has agents which are capable of performing rough cut capacity planning and generate Master schedule for a given period and also obtain relevant information about bill of materials and inventories status data. It has also agents to perform time phasing operation in order to generate list of items to be purchased, subcontracted and list of items to be manufactured in the given period. The output of this module will support purchasing and manufacturing planning activities such as process planning and scheduling.

1.2.5 Purchasing Activity Module (PAM)

Purchasing Activity Module (PAM) to execute all the activities involved in the purchases process right from material requisition submission to payment on purchase orders. This module utilize well designed item database as well a systematic quantitative procedure of vendor rating based not only on price of the item but on multiple performance factors.

1.2.6 Process Planning Module (PPM)

This module is concerned with generation of optimal process plan for any given components by employing individual agents to perform various functional elements related to process planning activities. Based on the out put of various agents that perform verities of functions related process planning the module finely presents the optimal process plan for the given part.

1.2.7 Production Scheduling Module (PSM)

Production Scheduling Module (PSM) consists of a set of agents which receive the necessary information form MRPM and PPM and determines optimal schedules for the manufacturing of the components required in a given period of time. The agents assigned to this module are capable of identifying standard times for any given work, perform capacity requirement planning, closely monitor shop floor activities etc.

1.2.8 Assembly and Testing Module (ATM)

This module consists of agents who maintain and provide information and ensure to carry out the assembly and testing of products smoothly and effectively.

1.2.9 Quality Control Module (QCM)

The process of assuring better quality for its products a manufacturing organization has to process large amount of data by employing suitable statistical and other methodologies. The agents of this module perform the activities that assure better quality of the products. They are capable of modelling various inspection and quality control approaches.

1.2.10 Customer Care and Sales Module (CSM)

The agents included in this module are capable of performing to all those activities related to the sales of the products as well as and after sales considerations. The activities related to packaging of the products as well as distribution of products to the expected locations are also included in this module.

1.2.11 Financial Activities Module (FAM)

The agents grouped under this module concentrate on flow of funds and other financial aspects such as investment decisions on both short term and long term activities. They maintain and update records of bills payables as well as bills receivables based on the transactions with the creditors (suppliers, service provides, subcontractors), debtors (distributors, customers).

1.2.12 Storage and Material supply module (SMM)

The agents grouped under this module continuously interact with purchasing module to share information regarding receipt of new purchase orders as well as supply of material to shop floor activities and accordingly update the available quantities of various materials. The updated information is made available for the functioning of agents of other modules such as MRPM and PAM.

1.2.13 Subcontracting Module (SUCM)

Subcontracting is a type of work contract that seeks to outsource certain types of work to other companies. Subcontracting or outsourcing strategy can provide not only cost and delivery benefits but also to introduce range of products. It is executed by identifying non-critical components of products and assigning the work to other organizations. The agents of this module collectively execute the processes necessary for assigning certain well defined work to a suitable subcontracting agency.

1.2.14 Transportation and Distribution Module (TDM)

One of the critical aspects of supply chain management is to find solution to the questions, where should the storage of products be located and how should they are moved from one supply chain location to another. Transportation and distribution activities are performed to take care of these aspects. The agents of his module are equipped with necessary information and procedure to carry out the expected function leading to effective transportation and distribution of goods to all categories of customers.

1.2.15 Distributors and Retailers module (DRM)

Distributors and retailers are companies that take products from producers and deliver to customers including general public. Formation of an efficient supply chain can be realized with a proper mechanism is created to effectively share the necessary information with the distributors and retailers. The agents developed for this module help to achieve this objective.

1.2.16 Supporting Databases

All the modules will have access to the common databases for their all their activities. The major databases include database for Industrial Engineering

Support (to provide all the necessary information related to Industrial Engineering tasks), Manufacturing Resource Data Base (to stores all the data related to manufacturing activities which are compiled using the syntax of artificial intelligence techniques) and also Knowledge Base.

Further, necessary agents have been developed for the following modules: MRPM, PAM, PPM and PSM i.e. for Manufacturing Resource Planning, purchasing process, Process Planning and scheduling activities to demonstrate the information flow with other organizations as well as the flow of information within the organization. The agents for the purchasing functions, which are developed with Visual Basic.net platform, are demonstrated in detail in the thesis.

1.3 OUTLINE OF THE THESES

The research work that has been carried out is described in detail in various chapters of this report. A brief description of chapters is given below.

The Chapter 1 is the current chapter which presents the need for carrying out this research activity along with an overview of the work and also objectives of the study.

A detailed literature survey has been conducted on various aspects of supply chain management, information sharing aspects as well as agent technology. The details of this survey are highlighted in the Chapter 2 along with the outcomes of the literature survey which lead to the problem identification of the current work.

In the Chapter 3 an overview of agent Technology and its fundamental concepts have been described that serve as building blocks for this research work.

The Chapter 4 presents a description about the various agents that haven been identified for carrying out the activities related to MRP and Process planning along with the related frame works. The ultimate objective is to present an overall picture if flow of information.

The Chapter 5 is used to present the basic procedure of purchasing process that is normally carried out in any organization and the various agents that perform all the activities of purchasing process automatically with efficient flow of information.

The Chapter 6 describes in detail the functioning of various agents developed for the purchasing process in order to demonstrate the flow of information and information sharing along the supply chain units.

The Chapter 7 presents a case study to describe the execution of a set of agents developed in the present work which perform activities related to material requirements planning and procurement of material.

The Chapter 8 is the last chapter which summarizes the work carried out, lists the special features of work along with merits and limitations and provides information about the scope for further work indicating the directions for future research activities that can be undertaken in the current topic.

Chapter 2

Literature Review

Global competition and rapidly changing customer requirements are forcing major changes in the production styles and configuration of manufacturing organizations. Manufacturing strategies are therefore shifting to support global competitiveness, new product innovation and rapid market responsiveness. For this manufacturing systems will need to satisfy the following fundamental requirements[8,13].

- *Enterprise Integration:* In order to support global competitiveness and rapid market responsiveness, an individual or collective manufacturing enterprise will have to be integrated with its related management systems and its partners via networks.

- *Distributed Organization:* For effective enterprise integration across distributed organizations, distributed knowledge-based systems will be needed.

- *Heterogeneous Environments:* Manufacturing systems will have to accommodate heterogeneous software and hardware in both their manufacturing and information environments.

- *Interoperability:* Heterogeneous information environments may use different programming languages, represent data with different representation languages and models, and operate in different computing platforms.

- *Open and Dynamic Structure:* It must be possible to dynamically integrate new subsystems (software, hardware, or manufacturing devices) into or remove existing subsystems from the system without stopping and reinitializing the working environment.

- *Cooperation:* Manufacturing enterprises will have to fully cooperate with their suppliers, partners, and customers for material supply, parts fabrication, final product commercialization, and so on. Such cooperation should be in an efficient and quick-response manner.

- *Integration of Humans with Software and Hardware:* People and computers need to be integrated to work collectively at various stages of the product development and even the whole product life cycle, with rapid access to required knowledge and information.

- *Agility:* Considerable attention must be given to reducing product cycle time to be able to respond to customer desires more quickly. Agile manufacturing is the ability to adapt quickly in a manufacturing environment of continuous and unanticipated change and thus is a key component in manufacturing strategies for global competition. To achieve agility, manufacturing facilities must be able to rapidly reconfigure and interact with heterogeneous systems and partners.

- *Scalability:* Scalability means that additional resources can be incorporated into the organization as required. This capability should be available at any working node in the system and at any level within the nodes. Expansion of resources should be possible without disrupting organizational links previously established.

The above mentioned facts indicate that present evolution of production systems and markets is forcing producers, distributors and vendors to integrate their operations into large-scale networks of different services for managing materials and products, information and capitals, i.e. into a 'supply chain'. 'Integration' has become the new paradigm in organizing business lines, and 'Supply Chain Management (SCM)' is becoming the related organization approach that can help in managing interactions among concurrent firms as well as markets[14].

2.1 HISTORY OF SCM

During the 1990s, many manufacturers and service providers sought to collaborate with their suppliers and upgrade their purchasing and supply management functions from a clerical role to an integral part of a new phenomenon known as supply chain management. Since this aspect of supply chain management primarily focuses on the purchasing and supply management functions of industrial buyers, it is classified elsewhere as the purchasing and supply perspective of supply chain management. Correspondingly, many wholesalers and retailers have also integrated their physical distribution and logistics functions into the transportation and logistics perspective of supply chain management to enhance competitive advantage[8]. Over the last 10 years, these two traditional supporting functions of corporate strategy evolved along separate paths and eventually merged into a holistic and strategic approach to operations, materials and logistics management commonly referred today as supply chain management[15,16].

Integrating the purchasing and logistics functions with other key corporate functions can create a closely linked set of manufacturing and distribution processes. It allows organizations to deliver products and services to both internal and external customers in a more timely and elective manner[17]. To further exploit the competitive advantage associated with integrated processes, some leading organizations adopt a strategic approach to managing the value chain, such as forming strategic alliances with suppliers and distributors instead of vertical integrating; inter-company competition is elevated to inter-supply chain competition. Although supply chain management developed along two separate paths, it has eventually merged into a unified body of literature with a common goal of waste elimination and increased efficiency[15].

The term 'supply chain' is only about 20 years old, having being coined by consultants in various parts of North America and Europe to explain the concept of managing an organization in the light of the activities, resources and strategies of the other organizations on which it relies. It is based on a rational model: the assumption is that, if a strategist in one firm (a customer) makes a plan for something to happen in another firm (a supplier), the desired activity will take place. This logic is further extended to include indirect relationships – the suppliers' suppliers. The chain is thus a chain of firms over which one firm expects to be able to exercise control[18,19].

The lack of a universal definition of supply chain management is in part due to the way the concept of supply chain has been developed[20]. Such a multidisciplinary origin and evolution is reflected in the lack of robust conceptual frameworks for the development of theory on supply chain management. As a consequence the schemes of interpretation of supply chain management are mostly partial or anecdotal with a relatively poor supply of empirically validated models explaining the scope and form of supply chain management, its costs and its benefits[21,22].

2.2 SUPPLY CHAIN

A supply chain is a network of business units that enables the collection of raw material, its transformation into products and the delivery of these products to consumers through a distribution system[2]. The supply chain of a manufacturing enterprise is a world-wide network of suppliers, factories, warehouses, distribution centers and retailers through which raw materials are acquired, transformed and delivered to customers Supply chain encompasses not only those activities involved in the flow and transformation of goods from the raw material stage to the finished product, but also those which are associated with information flows, cash flows and product flows in an organization. Generally, several companies are linked together in this process, each adding value to the product as it moves through the supply chain.

Most supply chains exhibit these basic characteristics:[2]. The supply chain includes all activities and processes to supply a product or service to a final customer.

1. Any number of companies can be linked in the supply chain.

2. A customer can be a supplier to another customer so the total chain can have a number of supplier customer relationships.

3. While the distribution system can be direct from supplier to customer, depending on the products and markets, it can contain a number of distributors such as wholesalers, warehouses, and retailers.

In its simplest form, a supply chain is composed of a company and the suppliers and customers of that company. The following is the basic group of participants that creates a simple supply chain[2]

- **Producers:** Producers or manufacturers are organizations that make a product. This includes companies that are producers of raw materials and companies that are producers of finished goods.

- **Distributors:** Distributors are companies that take inventory in bulk from producers and deliver a bundle of related product lines to customers. Distributors are also known as wholesalers.

- **Retailers:** Retailers stock inventory and sell in smaller quantities to the general public.

- **Customers:** Customers or consumers are any organization that purchases and uses a product.

- **Service Providers:** These are organizations that provide services to producers, distributors, retailers, and customers.

Recently, a new approach to the analysis of supply chains has been identified, which has proven to be of significant relevance to companies that have adopted it. This approach is based on the integration of different functions (e.g. purchasing, production, distribution, and storage) in the supply chain into a single optimization model. The basic idea behind this approach is to simultaneously optimize decision variables of different functions that have traditionally been optimized sequentially[23].

Companies in any supply chain must make decisions individually and collectively regarding their actions in five areas (some times referred as drivers). The sum of the decisions taken in these areas will define the capabilities and effectiveness of a company's supply chain[2] Effective supply chain management calls first for an understanding of each driver and how it operates. Each driver has the ability to directly affect the supply chain and enable certain capabilities.

a. *Production*—what products does the market want? How much of which products should be produced and by when? This activity includes the creation of master production schedules that take into account plant capacities, workload balancing, quality control, and equipment maintenance.

b. *Inventory*—what inventory should be stocked at each stage in a supply chain? How much inventory should be held as raw materials, semi finished, or finished goods? The primary purpose of inventory is to act as a buffer against uncertainty in the supply chain. However, holding inventory can be expensive, so what are the optimal inventory levels and reorder points?

c. **Location**—where should facilities for production and inventory storage be located? Where are the most cost efficient locations for production and for storage of inventory? Should existing facilities be used or new ones built? Once these decisions are made they determine the possible paths available for product to flow through for delivery to the final consumer.

d. **Transportation**—how should inventory be moved from one supply chain location to another? Air freight and truck delivery are generally fast and reliable but they are expensive. Shipping by sea or rail is much less expensive but usually involves longer transit times and more uncertainty. This uncertainty must be compensated for by stocking higher levels of inventory. When is it better to use which mode of transportation?

e. **Information**—how much data should be collected and how much information should be shared? Timely and accurate information holds the promise of better coordination and better decision making. With good information, people can make effective decisions about what to produce and how much, about where to locate inventory and how best to transport it.

Models have been developed for controlling supply chains based on the assumption that one organization can intervene in the business relationships of another. For the design of a supply chain as well as for a production system, the relationship between utilization, range, lead time and work in progress is very important. But a firm must be able to develop and implement its own strategy in order to ensure its survival and attract investment as nobody will invest in a firm that is controlled by someone else in the way that the supply chain management concept suggests[24,25].

2.3 SUPPLY CHAIN MANAGEMENT

Supply-chain management is the strategic, tactical, and operational decision making that optimizes supply-chain performance. The strategic level defines the supply chain network; that is, the selection of suppliers, transportation routes, manufacturing facilities, production levels, and warehouses etc (refer Fig. 2.1). The tactical level plans and schedules the supply chain to meet actual demand. The operational level executes plans. Tactical – and operational-level decision-making functions are distributed across the supply chain. The aim of supply chain management is to manage these activities so that products go through the business net-work in the shortest time and at the lowest possible costs. Improving supply chain management is a key strategy for increasing the enterprise's competitive position and profitability.

The development and evolution of supply chain management owes much to the purchasing, supply management, transportation and logistics literature. As such, the term supply chain management is used in many ways, but three distinct descriptions dominate prior literature. Firstly, supply chain management may be used as a handy synonym to describe the purchasing and supply activities of manufacturers. Secondly, it may be used to describe the transportation and logistics (and also reverse logistics) functions of the merchants and retailers[26]. Finally, it may be used to describe all the value-adding activities from the raw materials extractor to the end users, and including recycling. However, it should be no surprise that the various descriptions overlap in some cases[27].

The selection of a good supplier is critical to the financial success since supply chain management focuses on the development of cooperation and trusting relationships between supply chain partners. In the past, most companies often select supply chain suppliers by the price they offered. Therefore, companies often consider the lowest price as their choices for supply chain suppliers. In this competitive environment, in order to raise profits, reduce cost and strengthen the competition advantage, price is not the only reason for selecting suppliers. Many companies began to deliberate for cooperation. In addition, to analyse the suppliers' data for extracting the useful information, to assist in partner selection is critical in a successful SCM[12,28].

Enlightened suppliers realize that it is in their own interests for their customers to be successful. Once this is accepted, a partnership programme designed to aid success is an ideal strategy to pursue. By working in partnership, a smaller company purchasing from a much larger supplier can take advantage of resources and expertise that would be difficult to self-finance: as a result, it can offer extended services to its customers. The potential for partnership should be viewed not only as a one-to-one relationship between two companies, but as an opportunity for all elements of the supply chain to work together to eliminate waste. Competitive advantage will be achieved only through cooperation – not confrontation[29,30]. By encouraging partnerships, not only with their direct suppliers, but with their suppliers' suppliers, the benefits of this manner of conducting business can be enjoyed by the entire value chain[31]. Supply chains are increasingly susceptible to unplanned, unanticipated disruptions[32].

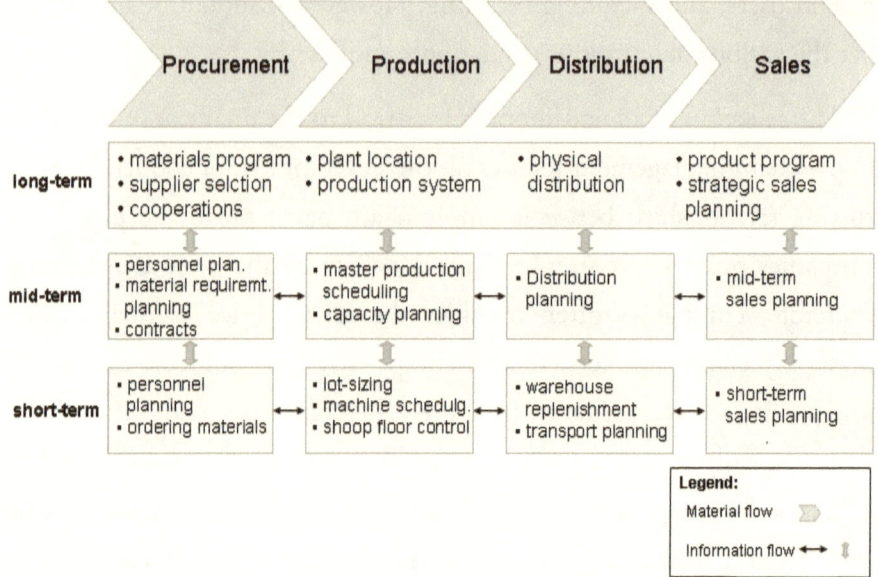

Fig. 2.1: Supply Chain Management-An Overview[33]

Research in the West over 15 years has concluded that it is not possible to manage supply chains or networks in the way that the extraordinary situation in Japan has made such techniques possible there[34]. Instead, organizations may only seek to manage within it to influence other firms, not manage them. This makes sense for a strategist to make a plan for ones own firm's activities and try to influence other firms (that is, either directly or indirectly) so that the plan may be complemented and therefore successful. Recent researches that have conducted in Japan indicate that the supply relationships in that country may be changing as a result of the severe recession that it is currently encountering. It appears that a less customer-dominated approach may be emerging in Japan – especially in international sourcing[19].

Supply chain management has exploded onto the business scene as one of corporate management's major concerns over the past decade. The reasons are clear. 70% of a firm's sales revenues are, on average, spent on supply chain-related activities from material purchases to the distribution and service of finished products to the final customer and hence need to be modeled[35]. As the world's economy becomes increasingly competitive, sustaining competitiveness and the resulting profitability depends less on the ability to raise prices. Instead, firms need to compete on the basis of product innovation, higher quality, and faster response times, all of which must be delivered, in most cases simultaneously and always at the lowest costs attainable[36]. These competitive dimensions cannot be delivered without an effectively managed supply chain. Firms with the most competitive supply chains are and will continue to be the big winners in contemporary business. Strategic supply management has the potential for significant value creation for the firm. Business professionals who have long been involved in supply management understand its power to create value. The emergence of e-procurement in the last few years is creating a higher profile for supply management, boosting its visibility to top management[37].

Supply Chain Management (SCM) has become of increasing interest as companies seek to improve customer delivery performance, reduce inventory and increase both flexibility and responsiveness. Many companies have experienced a growth in supply chain networks as a result of intense competition brought on by globalisation. Inbound supplies may come from countries where material and labor costs are cheaper. In addition, outgoing products often move through larger and broader distribution channels as companies strive for new markets and increased market share[38].

As global markets evolve, supply chain managers are faced with many new challenges, as traditional approaches to managing supply chains prove increasingly ineffective[39]. The integration of quality management principles offers potential for broadening the perspective of supply chain management from its traditional narrow focus on costs and competitive relationships to a focus on cooperative relationships between members of the supply chain and the strategic importance of supply chain management to the achievement of cumulative competitive capabilities[40].

In particular, Manufacturing enterprises face increasingly severe competition in both domestic and international markets. To improve, or at least to maintain, their enterprises' competitive position, manufacturing managers have to continue to improve their operations[41]. Research shows that activities such as the implementation of *Advanced Manufacturing Technology* (AMT) result in only marginal improvements if they are not accompanied by initiatives that improve linkages within the whole supply chain[118].

There is a stress on study about how the two strategic components, supplier involvement and manufacturing flexibility, can be integrated and, how such integration affects the performance of manufacturing flexibility[42]. The premise of this study is that both supplier involvement and manufacturing flexibility are multi-dimensional concepts; managers must understand how various supplier activities correspond to different dimensions of manufacturing flexibility and supply chain dynamics[43]. This knowledge enables firms to align their supply chain efforts with their manufacturing flexibility programs[44].

The literature characterizing environmental management within the supply chain has also been slowly building, but remains sparse. Moreover, investment by plants in environmental technologies cannot be made independently of other organisations in the supply chain[45]. Recoverable manufacturing systems minimize the environmental impact of industry by reusing materials and reducing energy use. Over the past decade or so, firms have faced increasing pressures from consumers and from governmental regulations to become more environmentally responsible. These systems are widespread and are profitable in addition to sustainable development[46].

Dynamic SCM requires integrated decision making amongst autonomous chain partners with effective decision information synchronization amongst them. By exploiting flexibility in supply chain structures, better performance can be achieved. Similarly, by judiciously employing decision flexibility and the associated dynamic control amongst autonomous supply chain nodes, many improvements are possible[13].

In SCM, a phenomenon called 'the bullwhip effect' has attracted considerable attention. The bullwhip effect refers the amplification of demand variability from a downstream site to an upstream site in the supply chain, and it has become the important index of supply chain performance in its operation process. The bullwhip effect is overwhelming, not only in supply chain operations, but also in modern logistics, even in the ERP system, e-commerce system, and other management systems. The devastating consequences caused by the bullwhip effect are obvious, such as a redundant inventory, excessive production and resultant costs, ineffective transportation and laggardly logistics, inefficient operations, and low economic benefits of supply chain system, which is the most important[13].

Moving one step ahead of supply chain management, among the managerial and organizational changes required in the face of global competition, virtual enterprise has emerged to adapt this competitive market environment[2]. Virtual enterprise is an organization created from physically distributed constituents, which are linked electronically to enable interaction and cooperation normally associated with a centralized enterprise[33,3]. The concept of virtual enterprise represents highest-level inter-enterprise integration.

2.4 INFORMATION TECHNOLOGY AND SCM

The requirement for organizations to become more responsive to the needs of customers, the changing conditions of competition and increasing levels of environmental turbulence is driving interest in the concept of "agility".

What it really means for an organization to be "agile", as opposed to just being efficient, effective, lean, customer-focused, able to add value, quality-driven, proactive rather than reactive, etc., has been the source of considerable debate and academic conjecture[47].

Hence, many companies are pursuing an agile enterprise that quickly senses and responds to customer preference changes, competitors' product innovations and advanced manufacturing technologies, economy status, exceptions, and interruptions in business. To streamline an agile manufacturing system of a global firm facing a high demand of market service, supply chain management (SCM) plays an important role. To remain competitive in today's business environment, it is important to effectively coordinate a wide variety of organizations by sharing necessary information. The exchange of timely and accurate information among business partners substantially improves the performance of the supply chain and resolves problems taking place due to lack of communications. Capturing customer preferences is essential to take a dominant position in competitive markets. As product customization is greater than ever, demand mix may vary widely with given similar total demand quantity. Manufacturers mass-producing multiple customizable products need strategies to deal with the increasing uncertainty of the demand mix[48].

In order to achieve the mass customization on a global scale, most competitive enterprises seek to enhance competitive performance by closely integrating the internal operations (Fig. 2.2) and by effectively linking them with the external operations of suppliers, customers (Fig. 2.3) and other business partners. As such, enterprises focus only their core capabilities on their value chain while they collaborate with other enterprises that have other complementary capabilities. This strategy requires the management of collaborative networks across business partners, and even business competitors.

As each enterprise operates as a node in the network composed of suppliers, customers, engineers and other specialized service providers, the collaborations among multiple business partners are becoming important[49].

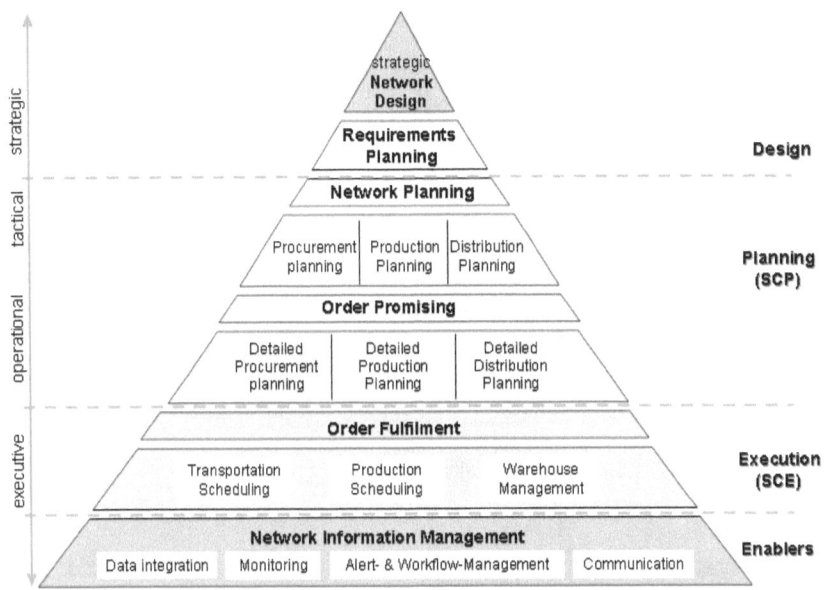

Fig. 2.2: Typical Information Technology for SCM[33]

There fore Supply chain management often requires the integration of inter – and intra-organizational relationships and coordination of different types of flows within the entire supply chain structure[50]. Inter-company integration and coordination via information technology has become a key to improved supply chain performance. Recent advances in information technology enable firms to manage effectively and inexpensively the coordination of not only the physical flow of materials but also the flow of different types of information such as demand, capacity, inventory, and scheduling, through an Internet enabled supply chain[51].

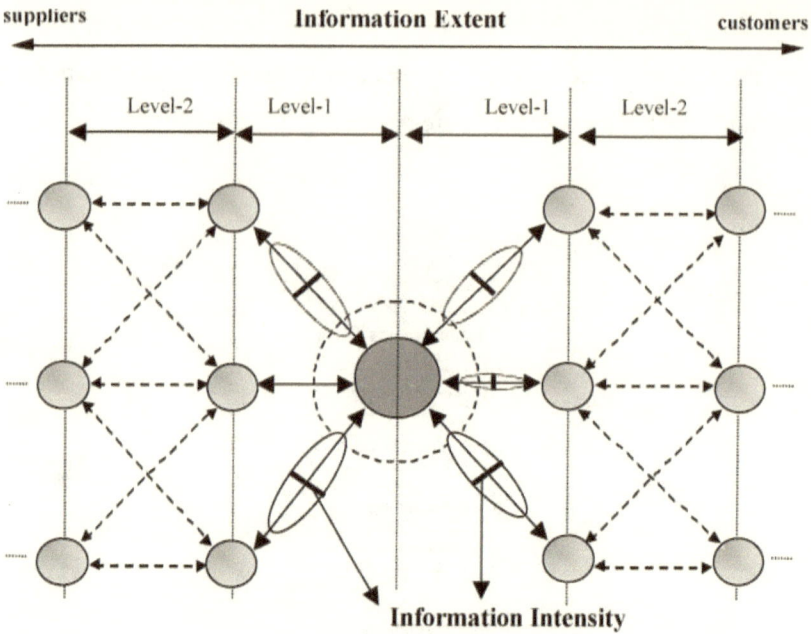

Fig. 2.3: Supply Chain and Information Coupling[52]

Inter organizational information systems are being implemented in all types of industries to establish electronic links between separately owned organizations. These may be simple electronic data interchange (EDI) systems for exchanging data such as purchase orders, advice of delivery notes and invoices or may involve more complex transactions such as integrated cash management systems or shared technical databases. The phenomenon of shared information systems is having a profound effect on business relationships, market structure and competition[53].

The development and implementation of inter organizational networks over the past decade has been accompanied by an expectation that fundamentally different organizational structures and forms of competitive behavior would begin to emerge. As the potential for linking the information systems of separate organizations has gradually been realized, parallel changes in the relationships between hitherto competing organizations have been suggested. In particular, these ideas emphasize the integrative potential of computer networks which promote information sharing, and the increasing collaboration which this implies. In other words, it is suggested that technological integration will prompt organizational integration[54,55].

It means that there is a need for a firm to restructure the relationships with their upstream and downstream firms along the entire supply chain in order to be able to achieve significant global performance improvements[50]. Nowadays, from manufacturers' point of view, customers become increasingly influential in terms of purchasing and bargaining power. In this connection, manufacturers need to cooperate or interact with suppliers to maximize the productivity at the minimum cost while satisfying customer requirements[56].

The deployment of the emerging concepts of information technology, strategic alliances, and business process re-engineering within the intra/inter-organizational context have become a popular prescription in enhancing supply chain management. In practice, effective methods for data communication between organizations are needed not only due to the potential economical benefits but also due to changes in legislation concerning traceability of raw materials and product lifecycle management in general. Both business process re-engineering (BPR) and the application of information technology (IT) are ingredients found in most of the popular prescriptions[57].

The introduction of information systems (IS) in supply chain management was originally limited to the automation of clerical functions. Information systems were viewed as providing infrastructural support to the value chain and they had an indirect impact on the competitiveness of a product[58]. Companies were able to save costs through information systems, but their use was not felt in a major way by customers. With intensification of competition, firms started to utilize information systems to influence directly the processes comprising the value chain. Through the utilization of such information Systems, companies have been able to integrate similar functions spread over different areas as well as curtail unnecessary activities, thus enhancing their capability to cope with the sophisticated needs of customers and meeting the quality standards[49].

The introduction and utilization of information systems can have a direct influence on value creation by integrating a firm's supply chain and, if utilized properly, can lead to higher quality products, enhanced productivity, efficient machine utilization, reduced space and, ultimately, increased logistics efficiency and flexibility by prototyping IS[59]. This means that IS utilization for routine data processing may be useful for cost reduction efforts in conjunction with the firm's strategy of cost leadership, but IS utilization for supply chain integration (SCI) can lead to differential and sustainable competitive advantage. Therefore, it is necessary to understand better how IS utilization in conjunction with SCI can lead to superior competitive advantage[22].

2.5 SCM AND INTERNET

The world of business is being changed to an e-economy by new forces of global competition, increased information availability, educated consumers, changing relationships, rapid innovations, and increasingly complex products. No industry is left untouched. In today's customer-focused marketplace, supply chain management has become a key to competitive advantage. But having accepted the challenge to create a synchronized supply chain that can compete in the future e-business and e-economy[60], the critical question is what concrete capabilities must then be mastered. We see an increase in the number and functionality of business models that use information systems that cross organizational boundaries, such as systems linking one or more firms with customers and/or suppliers. New business models emerge or old business models improve and experience a renaissance. But they all have a very short history and still have to prove their profitability and function[61].

Advances in information technology facilitate the deployment of electronic commerce within the supply chain system. Electronic data interchange (EDI) has been used to process business transactions between suppliers and customers since the early 1960s, covering various business activities such as sales/purchase, order processing, and the transfer of funds. In recent times, advanced inter organizational computer networks have enabled the application of new concepts in supply chain management, e.g., systems such as reversed inventory replenishment schemes.

Effective supply managers are aware that supplier development is also one of the new basics of supply management. So, the issue of a supplier's flexibility is generally dealt with and resolved before the e-procurement strategy is put in place. Linking of buyer and supplier not by inventory but by information referred as "virtual integration." The information link is essential to any truly comprehensive e-procurement strategy. The vehicle that facilitates virtual integration is the Internet.

The supply management and e-procurement literature is rich with estimates of the benefits of e-procurement. The potential is so great that e-procurement has turned the formerly looked-down-upon traditional purchasing function into a competitive weapon[62]. There is also an impact on a firm's asset base. Inventory levels can be significantly reduced. An effective e-procurement strategy where, for example, extranets link the systems of buyers and suppliers over the Internet, facilitates real-time exchange of information in the buyer's production schedule. The supplier can then adjust its output to meet the changes in the buyer's demand.

Of course, this implies that the supplier has developed capabilities allowing that degree of flexibility[63].

Forward-looking top executives are anxious to leverage the power of technology to improve the competitive position of their firms. At the same time, they are becoming increasingly aware of the power of effective supply management because it is at the input end of the supply chain that 50% or more of the firm's sales revenue is spent to help support company operations. It is the coming together of those two forces that has brought e-procurement center stage as the most significant development in supply management in recent years[47].

The rapid development of the internet has led to many changes in ways businesses are run. More and more companies are moving online for procuring raw materials, interacting with customers, and delivering directly to them by bypassing existing channels for product flow. Stores are turning into showrooms; computers are becoming the shopping store; and the shipment delivery boy is replacing the checkout counter. In a world where the customer has a lot of options and loyalty changes with the click of a button, achieving customer satisfaction will take much more than old-fashioned manufacturing shop floors and supply chains. We believe that achieving visibility of information is the first step towards success in this competition[64].

The Internet is starting to make world-wide web-based electronic commerce feasible, and the utilization of electronic commerce in supply chain management will increase in both sophistication and volume. The most significant benefit of the Internet for supply chain managers is in the provision of rapid, accurate and comprehensive information about each stage in their supply chain systems. In addition, the Internet has provided managers with the ability to be "agile" in managing their supply network.

This includes the ability to:[64]

1. Quickly adjust inventory levels,
2. Add or reduce carriers when needed,
3. Increase the speed in reacting to customer service problems,
4. More effectively manage distant facilities,
5. Reduce the level of paperwork in a supply chain system,
6. Adjust material throughput when necessary,
7. Track shipments more accurately,
8. Develop cost effective purchasing strategies,
9. Improve production scheduling and
10. Reduce operational redundancy in supply chain systems.

The increased flexibility in managing supply chains that the Internet has provided has enabled logistics managers to introduce "customization" and integrate customers more deeply into their supply chains. This is manifested in the ability of firms to customize service solutions for their customers when problems arise. For example, customers can trace their own shipments rather than rely on the carriers to do it for them. Vendors can view customer production schedules and inventory levels and coordinate their own supply chains with that of their buyers[65].

The impact of Internet technology on what has been traditionally called the purchasing process has been pervasive, starting with how suppliers and the internal members of the firm's buying team get involved with the specification development process to the systematic collection of data to completely and objectively evaluate supplier performance[66]. Given the potential that e-procurement holds for making a significant contribution to overall corporate strategy, it is incumbent on the supply manager to be able to make a business case for its adoption. Supply managers need to be comfortable speaking the language of top management by using financial measures to make their case.

Supply chain managers face interesting challenges as they decide which ways to integrate the Internet into their operations to enhance value[64]. The challenge those managers face is to select the mechanism that best fits each situation. The Internet is a standard computer communication network. Before its development, linking information systems across the supply chain was expensive and technically challenging. In this sense, the Internet empowers managers to extend the practice of supply chain management (SCM). The ability to integrate business activities with customers and suppliers leads to a competitive advantage, differentiating top performing firms. However, the Internet also enables managers to reduce acquisition costs by fostering price competition among suppliers using market mechanisms such as auctions[51].

Lancioni et al. surveyed 1000 US firms that were members of the Council of Logistics Management regarding their application of Internet technologies within their supply chains. They find that Internet adoption has increased from 1999 to 2001, moving away from indiscriminate use of Internet-related processes toward more focused, strategic applications and the development of precise and measurable goals. Their study, "Strategic Internet Trends in Supply Chain Management," shows that beyond cost reductions, the use of the Internet within the supply network increases productivity and profits for participating firms. The Internet allows firms to customize service solutions for their customers, which enhances the overall value and competitive position throughout the supply chain network[64].

The companies were eager to deploy the Internet wherever possible in their respective supply chain operations, e.g., in customer service, purchasing, inventory management, etc., hoping to reap some sort of benefits such as low costs and improved efficiency. This early adoption behavior may be classified as a "whatever sticks" approach, where the new Internet technologies were indiscriminately applied to discover which uses yielded the best results[66,18].

Inventory turnover can be monitored at customer demand locations and replenishment schedules established to reduce out-of-stock situations. JIT programs can be better managed through the Internet, thereby improving the payoffs from these hybrid operations[67]. Of course, the driving force that propels the growth of the Internet in supply chain management is the lower costs and improved profits that companies have enjoyed from its application. Money saved from supply chain efficiencies goes directly to the bottom line. This trend will continue to motivate companies to look for additional strategic applications of the Internet in their supply chain systems in the future[18].

Much has been said about how the business landscape has changed as a result of the Internet. Managers may use a variety of coordination mechanisms available through the Internet. However, for these to happen, management practices must be adapted to the new environment like e business[68]. The Internet facilitates the implementation of a variety of coordination mechanisms, such as information sharing, auctions, and electronic agents, for connecting members of the supply chain[69].

2.6 SCM AND E-MARKETPLACE

Supply chain management has been literally reinvented by the new networked technologies and the practices they facilitate, i.e. e-procurement, e-logistics, collaborative commerce, real-time demand forecasting, inventory management, true just-in-time (JIT) production, customer interface, and web-based package tracking[70]. By automating and streamlining the laborious routines of the purchasing function, purchasing professionals can focus more on strategic purchasing and corporate goal achievements. The goal of e-Procurement is not to make a supplier drop their prices or lower their margins but to achieve savings through management of material and administration costs[71]. Electronic marketplaces (e-marketplaces) have a profound influence on the way in which organizations manage their supply chains[72]. Given the importance of managing information flow with IT in the supply chain, it is critical that antecedents be identified[73]. Manufacturing companies have discovered that they either develop competitive strategies, tactics, and operations for the global market or be beaten by other manufacturers who have embraced more innovative approaches[74].

The unique feature of an Electronic Market (EM) is that it brings multiple buyers and sellers together (in a "virtual" sense) in one central market space. If it also enables them to buy and sell from each other at a dynamic price which is determined in accordance with the rules of the exchange, called an electronic exchange; otherwise it is called a portal. The important point, which differentiates an exchange from other B2B e-commerce companies, is that an exchange involves multiple buyers and sellers and it centralizes and matches buy and sell orders and provides post-trade information[75].

A business that offers goods or services for sale to other businesses, over the Internet, is not an EM even if it provides a price-setting mechanism that is normally associated with an EM, such as an auction, because there is only one seller. Beyond the perspective of supply chain management, the choice of an EM is depending on the different EM categories:

- Buyer-oriented, seller-oriented or neutral;
- Vertical or horizontal;
- Fix or variable pricing mechanism;
- Manufacturing or operating inputs; spot or system sourcing;

- Open or closed;
- Supported transactions phases;
- Aggregation or matching mechanism.

The type of the EM relationship transactional oriented, information-sharing, or collaborative oriented has an influence on the successful use of EMs within a supply chain agreement[61].

2.7 SCM AND MUTUAL TRUST WITH PARTNERS

The traditional adversarial, arms-length relationship is proving ineffective relative to the demonstrated benefits of true cooperative partnership. Such partnerships, however, can only be accomplished, when the present uncertainty between the manufacturer and the supplier is drastically reduced. This requires the development of effective signaling between parties and subsequent cooperative actions that instill mutual trust[38].

Managers who are serious about improving supply chain responsiveness should work towards building greater levels of trust with key-input suppliers, and explore opportunities for collocation and information sharing on a regular basis. The results also suggest that buyer–seller relationships may develop at two levels. At the industry level, intervening forces such as market power and legal contracts are closely related, yet appear to have little bearing on buyer–seller relationships at the interpersonal and cognitive level. Such a perspective is suggested by the result that the level of perceived buyer-dependence on a supplier was not associated with the level of trust in that supplier. In addition, when suppliers were willing to make site-specific asset commitments in the form of capacity and equipment, higher levels of trust were developed. As supply chain members are often separate and independent economic entities, a key issue in supply chain management is to develop mechanisms that can align their objectives and coordinate their activities so as to optimize system performance. Information is a keyword in the coordination.

The implication is important; even in cases when buyers do not have a large degree of control over a supplier, working with them to improve levels of trust may be helpful in improving supply chain responsiveness. As organizations seek to identify means of managing these new forms of relationships, researchers must develop new models and methods to identify which suppliers to approach in relationship development, the methods for implementing and sustaining such relationships, and the appropriate processes for dealing with conflicts within such relationships when they arise. Future studies should also consider the new elements developed in this study: site specific asset investments, human asset investment, contract formalization, dependence, and trust[76].

Of obvious importance is the extent of the *distribution* of the technology among suppliers. Whatever the distribution, several factors influence the ease with which the firm can gain access to and make use of the technology[77].

The *availability* of the technology defines the range of companies from which the technology can potentially be obtained. Its *stability* defines its stage of development and future prospects. Finally, the level of *dependence* defines the degree and type of reliance of the focal firm on the technology for competitive position. Whatever the external technology, the firm cannot by definition control it directly: instead, it must operate through its relationships with the suppliers who do[19].

Increasingly, the effort to manage technologies held by suppliers will form a central element of such strategies and will keep the purchasing function firmly in the technological front line. It was suggested that a consistent framework for evaluating external technologies and the suppliers that possess them can be a valuable tool in designing supplier relationships:

such a framework, based on the purchasing firm's technology strategy and the surrounding network structure will summarise the issues. Data from the case studies, while anecdotal, indicate that a variety of approaches to these issues are used in practice, based on a range of assumptions made by the firm regarding its position. Use of a framework would lead to several changes in the relationships through which identified key technologies are accessed[19].

It has now been recognized that small business is highly flexible and adaptable to change, be it environmental, operational or technological. These firms have been shown to penetrate wider-ranging markets, operate computer based manufacturing systems and possess sophisticated information systems. As strategic allies of large firms in meta-organizational contexts, subcontracting small and medium sized enterprises (SMEs) must follow suit in concepts such as just-in-time, environmental scanning and electronic trading. This implies a better utilization of EDI and a more equitable sharing of its benefits, contributing to greater integration and creative synergy within the network enterprise[78].

Changes take time, as they require adaptations not only in work procedures and behavior. It is also necessary that customer intentions in this respect coincide with those of the suppliers. Therefore it is understandable that EDI implementation takes time. It takes time to agree on standard file transfers, messages and codes. Furthermore, these changes need to be related to strategic ambitions with regard to supplier relationships and purchasing strategy. These issues need to go hand in hand[79].

2.8 AGENT TECHNOLOGY

An agent is autonomous, goal-oriented software that operates asynchronously, communicating and coordinating with other agents as needed[80,56]. It is a software entity that has a set of protocols which govern the operations of the manufacturing entity, a knowledge base, an inference mechanism and an explicit model of the problem to solve[81,82]. Agents communicate and negotiate with the other agents, perform the operations based on the local available information and may pursue their local goals. Agents possess sufficient knowledge and inferential capability and also possess sufficient authority to make commitments for users. However, the overall performance of the agent-based system has to be satisfactory even from a global point of view and depends primarily on the protocol that regulates the negotiations among the agents and on the quality of the data used to make these decisions[5].

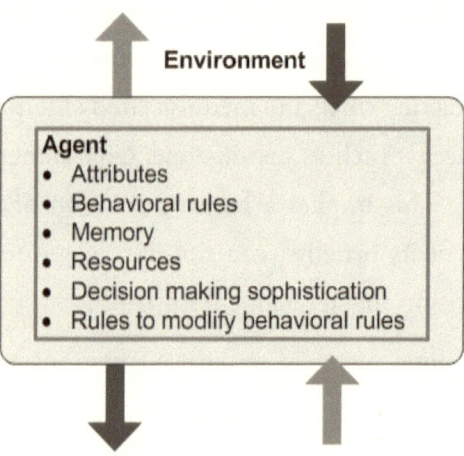

Fig. 2.4: Typical Agent[83]

This definition has both technical and organizational aspects. Technically, agents possess sufficient knowledge and inferential capability to behave in a manner that would be classified as "intelligent" if performed by a person.(refer Fig. 2.4) Organizationally, agents are entrusted with sufficient authority to make commitments for users. This enables them to represent their principals and adhere to the same corporate rules, policies and procedures required to be followed by people in the organization[83,84].

Agents have revolutionized manufacturing systems. In distributed intelligent manufacturing systems, the main function of agents is to integrate manufacturing enterprise activities such as design, planning, execution, simulation, distribution, forecasting between suppliers, customers and partners via a network. They are also used to represent various manufacturing sources like products, parts and operations to facilitate different manufacturing activities[85,86].

In the form of agent technology, in recent years new software architecture for managing the supply chain at the tactical and operational levels has emerged. It views the supply chain as composed of a set of intelligent (software) agents, each responsible for one or more activities in the supply chain and each interacting with other agents in planning and executing their responsibilities[7]. An agent is an autonomous, goal-oriented software process that operates asynchronously, communicating and coordinating with other agents as needed.

The common characteristics possessed by an agent are:[87,88]

- **Autonomy:** The agent is able to do at least part of its functionality independently and follow goals autonomously, that means without interactions or commands from the environment.
- **Intelligence:** The agent has some specialized knowledge in one or more application fields.
- **Interaction:** The agent is able to collect information or to react on conditions of its environment.
- **Reactivity:** An agent must be capable of reacting appropriately to inputs from its environment.
- **Pro-activity/Goal-orientation:** An agent does not just react to changes to its environment but it takes the initiative. To accomplish this property an agent must have well defined goals.

- **Learning:** An agent has to change its behavior based on its previous experience.

- **Mobility:** Mobility enables an agent to transport itself from one node of a network to another.

- **Communication/Cooperation:** An agent can use the communication capability to make contact with its environment which includes an interaction with other agents or human users Cooperation of several agents provides faster and better solutions for complex tasks since a single agent is often unable to accomplish complex tasks.

- **Character:** An agent could have a "personality" and an "emotional" state. This property is often used for agents representing a virtual person.

An agent need not to posses, all the above mentioned characteristics. Based on the subset of above mentioned characteristics possessed by an agent it can be groped into one of the following types of agents: Collaborative Agents, Interface Agents, Mobile Agents, Information/Internet Agents Reactive Software Agents and Hybrid Agents.[14]

Of these, Mobile agents have the capability to cross network boundaries and access other computers in networks such as wide area networks or even the World Wide Web (WWW)[89]. These agents do not transfer and process data on the other computer, but rather perform tasks and transfer results to the user. They represent the programs that can be initiated on a single host and then made to migrate from host to host over a network. At each host, a process can be spawned which will provide a "black-box" view into that host's information. This provides access to necessary information, while maintaining privacy for company sensitive information. Thus it can support the requirement for supply chain.

Agent technology provides a natural way to address the issues related to design and implement distributed intelligent manufacturing environments and provides software architecture for managing the supply chain. In distributed intelligent manufacturing systems, the main function of agents is to integrate manufacturing enterprise activities such as design, planning, execution, simulation, distribution, forecasting between suppliers, customers and partners. They are also used to represent various manufacturing sources like products, parts and operations to facilitate different manufacturing activities[87,60]. It views the supply chain as composed of a set of intelligent agents, each responsible for one or more activities in the supply chain and each interacting with other agents in planning and executing their responsibilities.

However, supply chains are really complex systems, and their management is a really complex task in which the cooperation of several intelligent agents together and their finalization to common industrial goals, need methods and procedures that are still either to be developed or, at the least, to be widely validated. The actual goal for an effective supply chain management is to obtain a good integration of all intelligent agents, such as to make each local strategy as cooperative as possible[90,91].

In today's manufacturing environment, the concept of producing the right product, at the right time, for the right price is a driving goal. To achieve this goal, a manufacturer must have great insight into the supply chain that feeds the manufacturing process. Many would advocate the use of MRP or ERP software systems to achieve this objective. The drawbacks with this approach range from an enormous initial investment to regimenting dated practices. Once such a system is put into place, it can stifle the move to new manufacturing technologies. On the other hand the flexibility of agent-based systems provides a great benefit over existing manufacturing control software[92].

2.9 SCM AND INFORMATION SHARING

Coordination plays a pivotal role in successful design and implementation of supply chains, especially for those that are formed by independent and autonomous companies[2] and hence, information sharing is a critical factor for successful business process management. One of the most effective ways to achieve the information sharing is to build an agent-based framework which models the dynamic structure of today's supply chain networks. Chan and Chan[5] analyzed the effects of negotiation-based information sharing in a distributed make-to-order manufacturing supply chain in a multi-period, multi-product type's environment, which is modeled as a multi-agent system. Information can only be exchanged through negotiation in the agent-based framework with delivery quantity and due date flexibility Four schemes, namely, stochastic model (STO), flexibility in delivery quantity and due date without information sharing (FLEX_NI), flexibility in delivery quantity and due date with partial information sharing (FLEX_PI), and flexibility in delivery quantity and due date with full information sharing (FLEX_FI), are considered. Simulation results indicate that FLEX_PI in the system has comparable performance in terms of total cost and fill rate against FLEX_FI.

Verdicchio and Colombetti[6] also stress that information sharing as a critical factor for successful business process management. A major problem facing manufacturing organisations is how to provide efficient and cost-effective responses to the unpredictable changes taking place in a global market. This problem is made difficult by the complexity of supply chain networks coupled with the complexity of individual manufacturing systems within supply chains. Current systems such as manufacturing execution systems (MES), supply chain management (SCM) systems and enterprise resource planning (ERP) systems do not provide adequate facilities for addressing this problem[93]. They argue that one of the most effective ways to achieve the information sharing is to build an agent-based framework which models the dynamic structure of today's supply chain networks.

Supply chain decisions are improved with access to global information. However, supply chain partners are frequently hesitant to provide full access to all the information within an enterprise[94]. A mechanism to make decisions based on global information without complete access to that information is required for improved supply chain decision making. Mobile agents can support this requirement and these are the programs that can be initiated on a single host and then migrate from host to host over a network. At each host, a process can be spawned which will provide a "black-box" view into that host's information. This provides access to necessary information, while maintaining privacy for company sensitive information. Mobile agents have been used for designing and managing the supply chains

Savarimuthu and Purvis[94] have described the architecture of agent-based workflow system that can be used for web service composition. In the context of an example from the apparel manufacturing industry, they have demonstrated how web Services can be composed and used. They suggest that with the advent of web services, more and more business organizations make their services available on the internet through web services and also use other services that are available on the corporate intranet.

Allwood and Lee have[91] proposed a new agent for the study of competitive supply chain network dynamics. The novel features of the agent include the ability to select between competing vendors, distribute orders preferentially among many customers, manage production and inventory, and determine price, based on competitive behavior. The structure of the agent is related to existing business models and sufficient details are provided to allow implementation.

Caridi et al[80] carried out a study of the Collaborative Planning Forecasting and Replenishment (CPFR) process for trading partners (belonging to the same supply chain) who are willing to collaborate in exchanging sales and order forecasts. The hurdles that arose in implementing CPFR in field applications indicate the need for providing collaboration process with an intelligent tool to optimize negotiation. To fulfill the need, two multi-agent models are proposed, according to different degrees of agents' capabilities. Through simulation studies it is indicates that the agents-driven negotiation process (by comparison with CPFR without intelligent agents) benefits in terms of costs, inventory level, stock-out level and sales.

The architecture proposed by George et al[95] addresses the issue of heterogeneous information management, the need for analyzing mixed manufacturing approaches, and efficiency and accuracy tradeoffs of analysis methods. The work proposes five distinct layers to address these requirements: an XML/RDF/DAML Information Access Layer, a Resource Vector Representation Layer, an Analysis Layer, a Coordination Layer, and a Visualization Layer. The proposed work is based on rich representation of manufacturing resources, and advanced analysis techniques.

Zhang et al[43] have presented an approach that would enable manufacturing organizations to dynamically and cost-effectively integrate, optimize, configure, simulate, restructure and control not only their own manufacturing systems but also their supply networks, in a coordinated manner to cope with the dynamic changes occurring in a global market. This is realized by a synergy of two emerging manufacturing concepts: Agent-based agile manufacturing systems and e-manufacturing. The concept is to represent a complex manufacturing system and its supply network with an agent-based modeling and simulation architecture and to dynamically generate alternative scenarios with respect to planning, scheduling, configuration and restructure of both the manufacturing system and its supply network based on the coordinated interactions amongst agents.

Macal and North[83] described Agent-based modeling and simulation as a new approach to modeling systems comprised of autonomous, interacting agents. They also presented a tutorial describing the theoretical and practical foundations of ABMS, identify toolkits and methods for developing ABMS models, and provide some thoughts on the relationship between ABMS and traditional modeling techniques. Multi-agent systems (MAS) perfectly suit the demands for global flexibility, co-operation and, at the same time, local autonomy and hence, offer new perspectives compared to conventional, centrally organized architectures in the scope of supply chain management. Besides the functional requirements like an integrated Supply chain scheduling, non-functional requirements like the reliability and the flexibility of the system are also met[96].

Better integration as well as coordination among the different nodes in the supply chain through use of information technology (refer Fig. 2.5) has helped to improve performance in these systems[6]. Generative models using agent-based objects are a very natural way for understanding and designing complex adaptive systems[97].

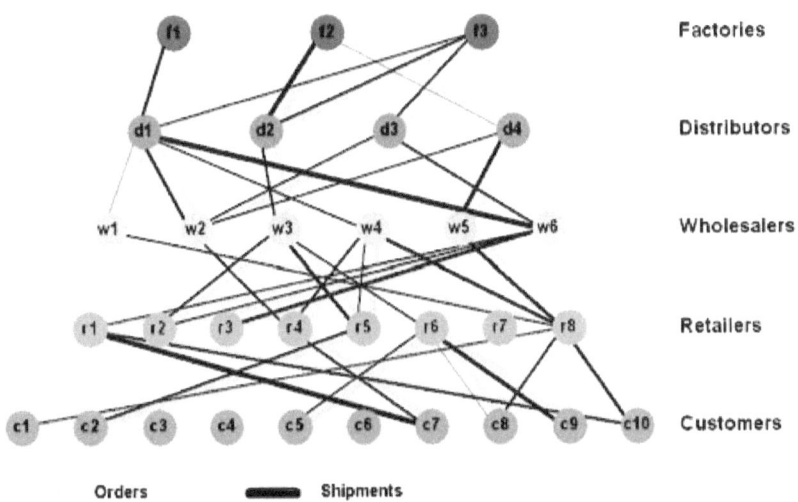

Fig. 2.5: Typical Supply Chain Network and Agents[83]

Some examples of agent based supply chains are, architecture of agent-based workflow system that can be used for Web Service composition[98], a new agent proposed for the study of competitive supply chain network dynamics[99] and study of the Collaborative Planning Forecasting and Replenishment (CPFR) process for trading partners (belonging to the same supply chain) who are willing to collaborate in exchanging sales and order forecasts[99]. The Supply Chain Agent Decision Aid System (SCADAS) proposed by Paolucci and Sacile serves as a tool to provide the flexibility of mobile agents while protecting company sensitive information[81]. And The Trading Agent Competition Supply Chain Management (TAC SCM) scenario[100] provides a unique test bed for studying and prototyping SCM agents by providing a competitive environment in which independently created agents can be tested against each other over the course of many simulations in an open academic setting

2.10 MULTI AGENT SYSTEMS

While agent technology supports the autonomy of supply chain partners[101], the Multi-agent systems (MAS) perfectly suit the demands for global flexibility, co-operation and, at the same time, local autonomy[80] and hence, offer new perspectives compared to conventional, centrally organized architectures in the scope of supply chain management. It addresses the architectural, the planning and the execution aspects of supply chain management, besides the functional requirements like an integrated SCM scheduling, non-functional requirements like the reliability and the flexibility of the system are also met[102].

A multi-agent system (MAS) is a system composed of multiple interacting intelligent agents. Multi-agent systems can be used to solve problems which are difficult or impossible for an individual agent or monolithic system to solve.

In the electronic business environment, as the supply chain management must deal with globalization, proliferating product variety, organizational barriers, and quick information sharing. Consequently, appropriate tools are needed to support supply chain management. The software agents are good candidates to overcome these challenges. The answer is a Multi-Agent System (MAS) to support Electronic Supply Chain Management (E-SCM). It may consist of a set of agents that are working together to maintain supplying, manufacturing, inventory and distributing[103]. The main operations of the software agents include:

1. Receiving information from customer orders.
2. Check the inventory.
3. Make the production schedule.
4. Issue the order of raw materials from the suppliers.
5. Receive the raw materials.
6. Production.
7. Deliver products to the customer.

In addition to the interface agents and communication protocols among agents.

A multi agent system views the supply chain as composed of a set of intelligent (software) agents, each responsible for one or more activities in the supply chain and each interacting with other agents in planning and executing their responsibilities[95].

MAS (Multi-agent System) technique, which is a branch of distribute artificial intelligence, has been regarded as one of the most promising approaches to solve scheduling problems under dynamic environments and has attracted a lot of attention recently[104,105], in particular to realize dynamic integration of scheduling algorithm.

Over the last decade the field of multiagent systems has evolved from its beginnings in distributed artificial intelligence as a discipline largely concerned with distributed problem solving, to the mature discipline we see today which brings together researchers from disciplines as diverse as economics, game theory, biology, robotics, software engineering, and artificial intelligence. This entire diversity make is difficult to provide students with the background needed for understanding multiagent approaches to problem-solving. This difficulty has been further compounded by the recent creation of sub-fields of study such as auction-based systems, robotic-based systems, software agents, and mechanism design approaches[106]. However, analyzing the benefits of multi-agent technology it is possible to conclude that it fulfils some of main requirements of the actual distributed manufacturing systems: autonomy (an agent can operate without the direct intervention of external entities, and has some kind of control over their behavior), cooperation (agents interact with other agents in order to achieve a common goal), reactivity and pro-activity (agents perceive their environment and respond adaptatively to changes that occur on it)[75].

The motivations for the increasing interest in MAS research include the ability of MAS to do the following[107]:

First is to solve problems that are too large for a centralized agent to solve because of resource limitations or the sheer risk of having one centralized system that could be a performance bottleneck or could fail at critical times. Second is to allow for the interconnection and interoperation of multiple existing legacy systems. Third is to provide solutions to problems that can naturally be regarded as a society of autonomous interacting components-agents. Fourth is to provide solutions that efficiently use information sources that are spatially distributed. Fifth is to provide solutions in situations where expertise is distributed.

Sixth is to enhance performance along the dimensions of (1) *computational efficiency* because concurrency of computation is exploited (as long as communication is kept minimal, for example, by transmitting high level information and results rather than low level data); (2) *reliability*, that is, graceful recovery of component failures, because agents with redundant capabilities or appropriate inter agent coordination are found dynamically (for example, taking up responsibilities of agents that fail); (3) *extensibility* because the number and the capabilities of agents working on a problem can be altered; (4) *robustness*, the system's ability to tolerate uncertainty, because suitable information is exchanged among agents; (5) *maintainability* because a system composed of multiple components-agents is easier to maintain because of its modularity; (6) *responsiveness* because modularity can handle anomalies locally, not propagate them to the whole system; (7) *flexibility* because agents with different abilities can adaptively organize to solve the current problem; and (8) *reuse* because functionally specific agents can be reused in different agent teams to solve different problems.

In MAS, agents are situated in an environment and they are able to "see" the environment through sensors and can possibly change the environment through their actions. The agents are autonomous that they are capable of making decisions based on their knowledge about the environment and/or other agents without the intervention of other agents. The agents exhibit social behavior, whereby they interact with other agents in order to achieve their goals. This section discusses three different types of agent architectures. The advantages and disadvantages of each type of agent are summarised. This is followed by looking at how, in an MAS, the agents can reach agreements. Particular attention is paid to the auction-based negotiation among agents.

Once agents are ready for collaboration, they will need to find the other agents they need to collaborate with. Such a task is easy if they know exactly which agents to contact and at which location. However, a static distribution of agents is very unlikely to exist: people are usually on the move and they are not always readily available to interact with others. The same holds true for dynamic multi-agent systems: agents need support to find other agents[108].

MASs using distributed control is commonly supposed to have several advantages over systems using alternative designs (e.g. central or hierarchical): they are robust, flexible, scalable, adaptable and amenable to mathematical analysis. Distributed control mechanisms for MAS are (1) biologically-inspired control for collaboration in a group of agents, in which local interactions among many simple agents lead to a desirable collective behavior, and (2) market-based control for resource allocation where agents compete for a limited resource whose capacity varies with time[109].

The production management system used by most of today's manufacturers consists of a set of separate application softwares, each for a different part of the planning, scheduling, and execution processes. For example, Capacity Analysis (CA) software determines a Master Production Schedule that sets long-term production targets. Enterprise Resource Planning (ERP) software generates material and resource plans. Scheduling software determines the sequence in which shop floor resources (people, machines, material, etc.) are used in producing different products. Manufacturing Execution System (MES) tracks real-time status of work in progress, enforces routing integrity, and reports labor/material claims. Most of these business applications are legacy systems developed over years. Although each of these software systems performs well for its designated tasks, they are not equipped to handle complex business scenarios, especially those which represent exceptions to the normal or expected business processes and whose resolution involve several applications. For example, consider the scenario involving a delay of the shipment date on a purchased part. This event may cause one of the following possible actions: (a) the manufacturing plan is still feasible, no action is required; (b) order substitute parts; (c) reschedule; or, (d) reallocate available material. To determine which of these actions to take, different applications (e.g., ERP and Scheduler) and possibly human decision-makers must be involved. Examples of other similar scenarios include a favorite customer's request to move ahead the delivery date for one of its orders, a machine breakdown being reported by MES, a crucial operation having its processing rate decreased from the normal rate, to mention just a few. Timely solutions to these and other scenarios are crucial to agile manufacturing. Unfortunately, current production management systems cannot support integrated solutions to such scenarios[110].

Unlike those stand-alone agents, agents in MAS collaborate with each other to achieve common goals. In other words, these agents *share* information, knowledge, and tasks among themselves. The intelligence of MAS is not only reflected by the expertise of individual agents but also exhibited by the emerged collective behavior beyond individual agents. From software engineering point of view, the approach of MAS is also proven to be an effective way to develop large distributed systems. Since agents are relatively independent pieces of software interacting with each other only through message-based inter-agent communication, system development, integration, and maintenance become easier and less costly. For instance, it is easy to add new agents into the agent system when needed. Also, the modification of legacy applications can be kept minimal when they are to be brought into the system.

2.11 SIGNIFICANCE OF PURCHASING IN SCM

The present day manufacturing companies face a global market characterized by numerous competitors, a steadily increasing complexity of business processes and a highly turbulent production environment. The purchasing function is central to the strategic operations of effective supply chain management[111]. This centrality is mainly due to the significant impact of material costs on profits, increased investments in inter-organizational advanced manufacturing and information technologies, and a growing emphasis on the just-in-time operations philosophy. The critical business processes of the purchasing function include supplier selection, negotiation of supply contracts, monitoring supplier performance, and acting as an interface between an organization and its suppliers[112].

The procurement function, traditionally long neglected as a clerical appendage of management, is gaining significant consideration. The trend towards purchasing complete systems is becoming the prominent tactic used by OEMs to increase their competitiveness[113]. The car manufacturers, on the one hand, see cost reduction as the main advantage of system sourcing with a possible bonus from value creation. Cost reductions in R&D and production, savings on inventory and storage space as well as value enhancements through concentrated know-how can be generated through collaborative effort[114,15]. Utilizing fewer direct suppliers reduces transaction costs while creating incentives to increase value through such potentials as R&D and supplier know-how. Suppliers, in turn, profit from longer contracts, higher sales volumes and stable relations which lead to an increase in competitiveness. System sourcing implies new organizational arrangements between car manufacturers and their suppliers to reflect a new form of cooperative business relations[115].

Supply chains are increasingly susceptible to unplanned, unanticipated disruptions. With the implementation of the practices of lean systems, total quality management (TQM), time-based competition and other supply chain improvement initiatives, managers now realize that their supply chains are fragile, particularly to environmental disruptions outside their control. a system is now emerging in purchasing to manage supply risk characterised as having a very low probability of occurrence, difficult to predict, and with a potentially catastrophic impact on the organization[116].

2.12 SIGNIFICANCE ISSUES ON INFORMATION SHARING

As highlighted above, a significant trend in the present market oriented manufacturing is large product variety and frequent design changes which translates into low production volumes[117]. Further, today's world market environment is rushing towards its total globalization, and is characterized by ever increasing pace in production and decreasing product cycle times. Therefore businesses that are more responsive to market changes and more sensitive to customers needs are more likely to survive and thrive in such environment[13]. The market demand for products with improved delivery performance for short and unpredictable life cycle, and the customer's demand for products tailored with their individual requirements have resulted in production in small batch-size and driven by customer's orders. Manufacturing strategies are therefore shifting to support global competitiveness, new product innovation and rapid market responsiveness[118,119]. Those companies that learn how to build and participate in strong supply chains will have a substantial competitive advantage in their markets. Based on these points the detailed literature survey has been presented in above sections.

The two major out come of this literature survey is that information sharing is most important requirement of efficient supply chain and multi agent modeling is most suitable for designing of supply chains. Further it is found that majority of the reported works deals with activities related to inventories and it is necessary to concentrate on those units of organization where manufacturing activities, in particular shop floor activities are involved which are in need of Agent technology. The approach should be from the point of view of Production Engineers. It is also felt that the advanced aspects of Agent technology are used only for information sharing and needs to be used for intelligence in manufacturing[120].

The previous discussions show that efficient and effective information flow along the supply chain is essential in order to improve its competitiveness in global market. Several studies have shown that there is a positive effect of information sharing along the supply chain one way to improve information flow is by sharing information among the business entities[79]. The available literature reveals that information sharing can enhance supply chain performance. Moreover, better integration as well as coordination among the different nodes in the supply chain through use of information technology has also helped improve performance in these systems. In the ideal case, a supply chain facilitates the availability of the right amount of the right product at the right price at the right place with minimal inventory across the network[121]. Despite its importance, little attention has been given in the literature to the issue of measuring the magnitude and the effectiveness of available information that logistics information systems provide[52].

Further the following are the key issues the must be considered in the analysis of flow of the information in a supply chain[122,126].

1. Every organization participating in a given supply chain should achieve enterprise integration, which means each unit of the organization will have access to information relevant to its task and will understand how its actions will impact other units of the organization thereby enabling it to choose alternatives that optimize the overall goals of the organization. Further to participate in a supply chain each organization should have open architectures for integrating its activities with those of the partners within wide supply chain networks.

2. As mentioned above, supply chain decisions are improved with access to global information. The more information about product supply, customer demand, market forecasts, and production schedules that companies share with each other, the more responsive everyone can be. However, supply chain partners are frequently hesitant to provide full access to all the information within an enterprise.

Each company has concerns about revealing information that could be used against it by a competitor. Hence, a mechanism to make decisions based on global information without complete access to that information is required for improved supply chain decision making environment. Within the supply chain companies should decide how much information should be shared with the other companies and how much information should be kept private.

3. Abundant, accurate information can enable very efficient operating decisions and better forecasts but the cost of building and installing systems to deliver this information can be very high. This has led to the development of innovative means to reduce supply chain cost within an individual company the trade-off between responsive-ness and efficiency involves weighing the benefits that good information can provide against the cost of acquiring that information.

4. Integration is challenging for the manufacturing enterprise because the manufacturing activities, including design, planning, simulating and tracking differ in format. A manufacturing enterprise consists of many different entities having varying degrees of scope and dimensions. Each entity has its own inputs, process control requirements, method s of optimization, transformation process and out puts.

This is due to the fact that although computerized applications in manufacturing are widespread, information processing is still fragmented since these computer applications have been developed without taking a general frame work into account and the hardware and software used are produced by a range of different firms. Hence there is a need to identify software development process which can support the management of heterogeneous or diversified information used for modeling a manufacturing enterprise.

5. Supply chain involves distributed intelligent manufacturing environment.

2.13 THE PROBLEM FORMULATION

To concentrate on the issues of information sharing as highlighted above, the current research activity has been carried out to develop a multi agent system model for supply chain which results in efficient sharing of information between various units of organization and also with other collaborating enterprises. The basic concepts of Agent based technology have been applied for the flow of information between various components of the supply chain formed for a manufacturing organization.

The following are the expected outcome of the proposed work:

1. Automation of purchasing process.
2. Frameworks for the automation of MRP, process planning and scheduling activities.
3. Overall frame work for various units of supply chain activities leading efficient flow of information.

Chapter 3

Concepts of Agent Based Approach

Agent-based Concept is one of the most vibrant and important areas of research and development to have emerged in information technology in recent years. Intelligent agent represents a new way of analyzing, designing and implementing complex software system. For agent-based technologies, the objectives are to create systems situated in dynamic and open environments, able to adapt to these environments and capable of incorporating autonomous and self-interested components. Agent-based systems provides concrete advantages such as: improving operational robustness with intelligent failure recovery, reducing sourcing costs by computing the most beneficial acquisition policies in online market and improving efficiency of manufacturing processes in dynamic environments. In particular, the characteristic of dynamic in which heterogonous systems must interact, span organizational boundaries and operate effectively within rapidly changing circumstances and with dramatically increasing quantities of available information is very significant.

3.1 DESCRIPTION OF AGENT

One aspect of agents that is broadly mentioned in the literature is the notion of agents as interactive entities that exist as part of an environment shared with other agents. This definition of an agent is taken from descriptions given by several authors, who describe agents as conceptual entities that perceive and act in a proactive or reactive manner within an environment where other agents exist and interact with each other based on shared knowledge of communication and representation.

An agent is simply another kind of software abstraction, an abstraction in the same way that methods, functions, and objects are software abstractions. Agents and objects share many characteristics; this sometimes makes it hard to differentiate between them. For example, Agent-Oriented Programming (AOP) could be considered a specialization of the Object-Oriented Programming (OOP) paradigm. OOP views systems as consisting of objects communicating with one another to perform internal computations, whereas AOP specializes this view to have agents (instead of objects), whose internal computations are based on beliefs, capabilities, and choices, that communicate with each other using messages adopted from speech-act theory. Although this view allows one to appreciate the similarities between agent and objects, their differences are less obvious.

With the agent-based approach, we can implement agents with sophisticated intellectual capabilities such as the ability to reason, learn, or plan. In addition, intelligent software agents can utilize extensive amounts of knowledge about their problem domain. This means that the underlying agent architecture must support sophisticated reasoning, learning, planning, and knowledge representation. The emerging paradigm of agent-based computation has revolutionized the building of intelligent and decentralized systems. The new technologies met well the requirements in all domains of manufacturing where problems of uncertainty and temporal dynamics, information sharing and distributed operation, or coordination and cooperation of autonomous entities had to be tackled.

Agent-based computation is a new paradigm of information and communication technology that largely shapes and, at the same time, provides supporting technology to the above trends. Agent theories and applications have appeared in many scientific and engineering disciplines. Agents address autonomy and complexity: they are adaptive to changes and disruptions, exhibit intelligence and are distributed in nature. In this setting computation is a kind of social activity. Agents can help in self-recovery, and react to real-time perturbations. Agents are vital in the globalization context, as globalization refers to an inherently distributed world both from geographical and information processing perspectives.

Agents – and similar concepts – were welcome in manufacturing because they helped to realize important properties as autonomy, responsiveness, redundancy, distributed ness, and openness. Agents could be designed to work with uncertain and/or incomplete information and knowledge. Hence, many tasks related to manufacturing – from engineering design to supply chain management – could be conducted by agents, small and large, simple and sophisticated, fine – and coarse-grained that were enabled and empowered to communicate and cooperate with each other.

The theory of computational agents goes back at least a quarter of a century when research in distributed artificial intelligence (DAI) had been initiated. In the early 90's the notion of agents appeared simultaneously also in information and communication technology (mobile, interface and information agents). Agents made the real breakthrough a decade ago or so when the emphasis in the mainstream AI research shifted: the focus on logic was extended and attention changed from goal-seeking to rational behavior; from ideal to resource-bound reasoning; from capturing expertise in narrow domains to re-usable and sharable knowledge repositories; from the single to multiple cognitive entities acting in communities.

These developments also coincided with the evolution of network-based computing technology, the internet, mobile computing, the ubiquity of computing as well as novel, human-oriented software engineering methodologies. All these achievements led to what is considered now the agent paradigm of computing. While this novel paradigm has several roots as far as theory, enabling technologies and applications are concerned, there is a general consensus about its two main abstractions:

- An agent is a computational system that is situated in a dynamic environment and is capable of exhibiting autonomous and intelligent behavior.
- An agent may have an environment that includes other agents. The community of interacting agents, as a whole, operates as a multi-agent system.

3.2 PROPERTIES OF AN AGENT

The most important common properties of computational agents are as follows:

- Agents act on behalf of their designer or the user they represent in order to meet a particular purpose.
- Agents are autonomous in the sense that they control both their internal state and behavior in the environment.
- Agents exhibit some kind of intelligence, from applying fixed rules to reasoning, planning and learning capabilities.
- Agents interact with their environment, and in a community, with other agents.
- Agents are ideally adaptive, i.e., capable of tailoring their behavior to the changes of the environment without the intervention of their designer.

Further agent properties, characteristic in particular domains and applications are mobility (when an agent can transport itself to another environment to access remote resources or to meet other agents), genuineness (when it does not falsify its identity), transparency, and credibility or trustworthiness (when it does not communicate false information willfully). Even though they exhibit only some of the above properties, agents relax several strong assumptions of classical computational intelligence: they typically have incomplete and inconsistent knowledge as well as limited reasoning capabilities and resources. Agents are individual problem-solvers with some capability of sensing and acting upon their environment, for deciding their own course of action, as well as for communicating with other agents. Depending on the actual problem and available technology at hand, agents can apply various faculties of problem solving, including searching, reasoning, planning, and learning. The notion of agents has a strong synthesizing power; hence the applied techniques may include both symbolic and sub-symbolic methods, classical and quantitative decision theory, as well as knowledge-based reasoning and sophisticated belief-desire-intention (BDI) models.

Eventually, a software agent is autonomous (and in particular have its own resources); is made in order to realize a set of goals" by acting in an environment (in function of its resources, its competences, its perceptions, the communications it gets.); have to reason, which enables it to select the good action, i.e. have sort of a processing unit (e.g. it is able to analyze its success and errors, it can learn and adapt itself.); can interact with other agents and can act for another agent, if it is intelligent and if the other has grant the right.

An agent evolves into an environment. It percepts some parameters of this environment, and with this information, it can update the partial representation he has of its environment (or not, if it is kind of a state-less agent). So generally, the agent has software components named sensors, which enable it to percept some things of the environment. Once the environment is updated, the agent can call its internal mechanic. It normally has a software component, supplied by a database (its memory") or not, which is its core algorithm, which will process the information, in order to return the appropriate action to do, considering the purpose, or set of purpose, it has.

3.3 GENERIC AGENT

The definition of "agent" is widely debated in the research community today. Agent is a software entity which functions continuously and autonomously in a particular environment and able to carry out activities in a flexible and intelligent manner that is responsive to changes in the environment. Ideally, an agent that functions continuously in an environment over a long period of time would be able to learn from its experience. An agent that inhabits an environment with other agents and processes to be able to communicate and cooperate with them, perhaps move from place to place in doing so. In general, agents must be autonomous, able to execute without user intervention.

They must be able to communicate with other software or human agents and to perceive the environment in which they reside. Agents can communicate with each other and with other systems and applications through business process interfaces. Agents can also perceive and monitor the environment in which they reside and make appropriate behavior choices; and, they are autonomous, able to execute without user intervention. Unlike humans, agents never get tired, do not make mistakes, can pay attention to every detail in massive amounts of data and work at electronic speeds.

Agents can play many roles and are well-suited for a wide variety of processes and applications in today's business world. They are particularly well-suited to areas such as:

- Process and workflow automation.
- Electronic commerce.
- Distributed problem solving.
- Internet applications.

Any software that behaves in an agent-like manner and exhibits one or more of the characteristics identified in the definition in the previous section is considered an agent. "Intelligence" in agents adds a more realistic, dynamic dimension to the characteristics of software agents. To be intelligent, agents must be able to work together on solving problems in a dynamic environment and must be able to communicate understandable results back to the user. As demonstrated in the scenario in the previous section, intelligent agents must be proactive and able to react to changing situations. Emulating more realistic behavior in day-to-day operations, intelligent agents can do much more than just match situational patterns or apply a static set of rules to solve a problem. An intelligent agent can operate in real time and use natural language to communicate; and, it is able to learn from the environment and be adaptive to user behavior.

All software agents are programs, but not all programs are agents.

The power of agent systems (software applications constructed in part or whole by agents) in the enterprise is in their ability to be experts individually but to collectively perform very complex activities through communication and coordination. This is not to say that only agent systems can do complex tasks, but, rather, that the agent-oriented approach makes it very logical and understandable to design, which generally translates to better code, less complexity, and greater reliability. From a software-engineering point of view, it is easier to build something reliable that knows one aspect of the problem than it is to build something that knows every aspect of the problem. Like any engineering problem, the features of the problem space and your requirements will dictate the most appropriate solution to the problem. Though very flexible and powerful, agents alone are not a silver bullet for a successful system. Intelligent agent applications do however enable secure, dynamic collaboration over complex business processes that deal with massive amounts of information to levels of accuracy, timeliness and quality never possible before.

Researchers and software companies have set high hopes on these so-called *software agents*, which "know" users' interests and can act autonomously on their behalf. Instead of exercising complete control, people will be engaged in a cooperative process in which both human and computer agents initiate communications, monitor events and perform tasks to meet users' goals. In essence, we need software agents because:

- More and more everyday tasks are computer-based.
- The world is in a midst of an information revolution, resulting in vast amounts of dynamic and unstructured information.
- Increasingly more users are untrained.

And, therefore, users require agents to assist them in order to understand the technically complex world we are in the process of creating. The number and type of application domains in which agent technologies are being applied to or investigated include workflow management, network management, air-traffic control, business process re-engineering, data mining, information retrieval/management, electronic commerce, education, personal digital assistants (PDAs), scheduling/diary management, etc.

In the enterprise, agents can be used for collaborative filtering and fusing of data, gathering and analyzing logical sets of data in real time, and then visualizing those large data sets at a glance. Intelligent agents can also monitor user behavior to automate repeated tasks. Effective uses of intelligent agents as they are applied in the enterprise include data mining, profile management, privacy management, rules management, and application management. Depending on the application, these systems can be rather simple or very complex. A simple example of an intelligent enterprise application would be a facilities management program where sensors monitor the environment and actuators make adjustments according to business rules, time of day, etc. More complex examples of intelligent applications would include logistical and supply chain planning operations. In such applications, the system reasons about the environment and builds optimized plans for resource allocation, routes and action sequences. It continuously evaluates and adjusts during execution. While being data intensive, intelligence in this domain can provide significant cost savings.

3.4 SIGNIFICANCE OF SOFTWARE AGENTS

While the original work on agents was instigated by researchers' intent on studying computational models of distributed intelligence, a new wave of interest has been fueled by two additional concerns of a practical nature: 1) simplifying the complexities of distributed computing and 2) overcoming the limitations of current user interface approaches. Both of these can essentially be seen as a continuation of the trend toward greater abstraction of interfaces to computing services. On the one hand, there is a desire to further abstract the details of hardware, software, and communication patterns by replacing today's program to – program interfaces with more powerful, general, and uniform agent-to-agent interfaces; on the other hand there is a desire to further abstract the details of the human-to-program interface by delegating to agents the details of specifying and carrying out complex tasks.

While there is little disagreement that future computing environments will consist of distributed software systems running on multiple heterogeneous platforms, many of today's most common configurations are, for all intents and purposes, disjoint: they do not really communicate or cooperate except in very basic ways (e.g., file transfer, print servers, database queries). The current ubiquity of the Web makes it easy to forget that until the last few years, computer systems that *could* communicate typically relied on proprietary or *ad hoc* interfaces for their particular connection. A shift from the network operating system to Internet and intranet-based network computing is taking place. As this transition takes place, we are seeing the proliferation of operating system-independent interoperable network services such as naming, directory, and security.

The concept of an agent has found currency in a diverse range of sub-disciplines of information technology, including computer networks, software engineering, artificial intelligence, human-computer interaction, distributed and concurrent systems, mobile systems, telematics, computer-supported cooperative work, control systems, decision support, information retrieval and management, and electronic commerce. In practical developments, web services, for example, now offer fundamentally new ways of doing business through a set of standardised tools, and support a service-oriented view of distinct and independent software components interacting to provide valuable functionality. In the context of such developments, agent technologies have increasingly come to the foreground. Because of its horizontal nature, it is likely that the successful adoption of agent technology will have a profound, long-term impact both on the competitiveness and viability of IT industries, and on the way in which future computer systems will be conceptualised and implemented.

3.5 AGENTS AS DESIGN METAPHOR

Agents provide software designers and developers with a way of structuring an application around autonomous, communicative components, and lead to the construction of software tools and infrastructure to support the design metaphor. In this sense, they offer a new and often more appropriate route to the development of complex computational systems, especially in open and dynamic environments. In order to support this view of systems development, particular tools and techniques need to be introduced. For example, methodologies to guide analysis and design are required, agent architectures are needed for the design of individual software components, tools and abstractions are required to enable developers to deal with the complexity of implemented systems, and supporting infrastructure (embracing other relevant, widely used technologies, such as web services) must be integrated.

3.6 AGENTS AS SOURCE OF TECHNOLOGY

Agent technologies span a range of specific techniques and algorithms for dealing with interactions in dynamic, open environments. These address issues such as balancing reaction and deliberation in individual agent architectures, learning from and about other agents in the environment, eliciting and acting upon user preferences, finding ways to negotiate and cooperate with other agents, and developing appropriate means of forming and managing coalitions (and other organisations). Moreover, the adoption of agent-based approaches is increasingly influential in other domains. For example, multi-agent systems are already providing new and more effective methods of resource allocation in complex environments than previous approaches.

3.7 AGENTS FOR SIMULATION

The use of agent systems to simulate real-world domains may provide answers to complex physical or social problems that would otherwise be unobtainable due to the complexity involved, as in the modelling of the impact of climate change on biological populations, or modelling the impact of public policy options on social or economic behavior. Agent based simulation spans: social structures and institutions to develop plausible explanations of observed phenomena, to help in the design of organisational structures, and to inform policy or managerial decisions; physical systems, including intelligent buildings, traffic systems and biological populations; and software systems of all types, currently including e commerce and information management systems.

3.8 TRUST AND REPUTATION

Many applications involving multiple individuals or organisations must take into account the relationships (explicit or implicit) between participants. Furthermore, individual agents may also need to be aware of these relationships in order to make appropriate decisions. The field of trust, reputation and social structure seeks to capture human notions such as trust, reputation, dependence, obligations, permissions, norms, institutions and other social structures in electronic form.

By modelling these notions, engineers can borrow strategies commonly used by humans to resolve conflicts that arise when creating distributed applications, such as regulating the actions of large populations of agents using financial disincentives for breaking social rules or devising market mechanisms that are proof against certain types of malicious manipulation. The theories are often based on insights from different domains including economics (market-based approaches), other social sciences (social laws, social power) or mathematics (game theory and mechanism design).

The complementary aspect of this social perspective relating to reputation and norms is a traditional concern with security.

Although currently deployed agent applications often provide good security, when considering agents autonomously acting on behalf of their owner several additional factors need to be addressed. In particular, collaboration of any kind, especially in situations in which computers act on behalf of users or organisations, will only succeed if there is trust. Ensuring this trust requires, for example, the use of: reputation mechanisms to assess prior behaviour; norms (or social rules) and the enforcement of sanctions; and electronic contracts to represent agreements. Whereas assurance deals primarily with system integrity, security addresses protection from malicious entities: preventing would-be attackers from exploiting self-organisation mechanisms that alter system structure and behaviour. In addition, to verify component sources, a self-organising software system must protect its core from attacks.

Various well-studied security mechanisms are available, such as strong encryption to ensure confidentiality and authenticity of messages related to self-organisation. However, the frameworks within which such mechanisms can be effectively applied in self-organising systems still require considerable further research. In addition, the results of applying self-organisation and emergence approaches over long time periods lead to concerns about the privacy and trustworthiness of such systems and the data they hold. The areas of security, privacy and trust are critical components for the next stages of research and deployment of open distributed systems and as a result of self organising systems. New approaches are required to take into account both social and technical aspects of this issue to drive the proliferation of self-organising software in a large range of application domains.

3.9 INTERACTION LEVEL COORDINATION

Coordination is defined in many ways but in its simplest form it refers to ensuring that the actions of independent actors (agents) in an environment are coherent in some way. The challenge therefore is to identify mechanisms that allow agents to coordinate their actions automatically without the need for human supervision, a requirement found in a wide variety of real applications. In turn, cooperation refers to coordination with a common goal in mind. Research to date has identified a huge range of different types of coordination and cooperation mechanisms, ranging from emergent cooperation (which can arise without any explicit communication between agents), coordination protocols (which structure interactions to reach decisions) and coordination media (or distributed data stores that enable asynchronous communication of goals, objectives or other useful data), to distributed planning (which takes into account possible and likely actions of agents in the domain).

3.10 NEGOTIATION

Goal-driven agents in a multi-agent society typically have conflicting goals; in other words, not all agents may be able to satisfy their respective goals simultaneously. This may occur, for example, with regard to contested resources or with multiple demands on an agent's time and attention. In such circumstances, agents will need to enter into negotiations with each other to resolve conflicts. Accordingly, considerable effort has been devoted to negotiation protocols, resource-allocation methods, and optimal division procedures. This work has drawn on ideas from computer science and artificial intelligence on the one hand, and the socio-economic sciences on the other. For example, a typical objective in multi-agent resource allocation is to find an allocation that is optimal with respect to a suitable metric that depends, in one way or another, on the preferences of the individual agents in the system.

Many concepts studied in social choice theory can be utilised to assess the quality of resource allocations. Of particular importance are concepts such as envy-freeness and equitability that can be used to model fairness considerations. These concepts are relevant to a wide range of applications. A good example is the work on the fair and efficient exploitation of Earth Observation. The centralised approach has the advantage of requiring only comparatively simple communication protocols. Furthermore, recent advances in the design of powerful algorithms for combinatorial auctions have had a strong impact on the research community. A new challenge in the field of multi-agent resource allocation is to transfer these techniques to distributed resource allocation frameworks, which are not only important in cases where it may be difficult to find an agent that could take on the role of the auctioneer (for instance, in view of its computational capabilities or its trustworthiness), but which also provide a test-bed for a wide range of agent-based techniques. To reach its full potential, distributed resource allocation requires further fundamental research into agent interaction protocols, negotiation strategies, formal (e.g. complexity-theoretic) properties of resource allocation frameworks, and distributed algorithm design, as well as a new perspective on what "optimal" means in a distributed setting.

Other negotiation techniques are also likely to become increasingly prevalent. For example, one-to-one negotiation, or bargaining, over multiple parameters or attributes to establish service-level agreements between service providers and service consumers will be key in future service-oriented computing environments. In addition to approaches drawn from economics and social choice theory in political science, recent efforts in argumentation based negotiation have drawn on ideas from the philosophy of argument and the psychology of persuasion. These efforts potentially provide a means to enable niches of deeper interactions between agents than do the relatively simpler protocols of economic auction and negotiation mechanisms. Considerable research and development efforts will be needed to create computational mechanisms and strategies for such interactions, and this is likely to be an important focus of agent systems research in the next decade.

3.11 COMMUNICATION

Agent communication is the study of how two or more software entities may communicate with each other. The research issues in the domain are long-standing and deep. One challenge is the difficulty of assigning meaning to utterances, since the precise meaning of a statement depends upon: the context in which it is uttered; its position in a sequence of previous utterances; the nature of the statement (for example, a proposition, a commitment to undertake some action, a request, etc); the objects referred to in the statement (such as a real world object, a mental state, a future world-state, etc); and the identity of the speaker and of the intended hearers. Another challenge, perhaps insurmountable, is semantic verification: how to verify that an agent means what it says when it makes an utterance.

In an open agent system, one agent is not normally able to view the internal code of another agent in order to verify an utterance by the latter; even if this were possible, a sufficiently-clever agent could always simulate any desired mental state when inspected by another agent. Key to this area is the need to map the relevant theories in the domain, and to develop a unifying framework for them. In particular, a formal theory of agent languages and protocols is necessary, so as to be able to study language and protocol properties comprehensively, and to rigorously compare one language or protocol with another. In addition, progress towards understanding the applicability of different agent communication languages, content languages and protocols in different application domains is necessary for wider adoption of research findings.

3.12 AGENT BASED MANUFACTURING

Agent based or Holonic manufacturing systems support a more plug-and-play approach to configuring and operating manufacturing processes, and thereby address increasing efforts to meet the needs for market responsiveness and mass customised products. A manufacturing Holon is an autonomous and cooperative building block of a manufacturing system for transforming, transporting, storing physical and information objects. It is intended to enable a "plug and play" approach to designing and operating a manufacturing system. In the last ten years, an increasing amount of research has been devoted to holonic manufacturing over a diverse range of both theoretical issues and industrial applications. Holon is a combination of the Greek word *holos* meaning whole and the Greek suffix *on* meaning particle or part as in proton or neutron.

It consists of a control part and an optional physical processing part. Hence, for example, a suitable combination of a machine tool, an NC controller and an operator interacting via a suitable interface could form a Holon which transforms physical objects in a manufacturing environment. Other examples of manufacturing holons could be products and their associated production plans, customer orders and information-processing functions. A Holon can itself also consist of other holons which provide the necessary processing, information, and human interfaces to the outside world. A "system of holons which can cooperate to achieve a goal or objective" is then called a holarchy. Holarchies can be created and dissolved dynamically depending on the current needs of the manufacturing process. Hence, the intention is that a combination of different holons will be responsible for the entire production operation, including not only the production planning and control functions, but also the physical transformation of raw materials into products, the management of the equipment performing the production tasks, and the necessary reporting functions. In this case the set of holons is referred to as a holonic manufacturing system. A holonic-systems view of the manufacturing operation is one of creating a working manufacturing environment from the bottom up. By providing the facilities within holons to both (a) support all production and control functions required to complete production tasks and (b) manage the underlying equipment and systems, a complete production systems is built up like a jigsaw puzzle!

Manufacturing operations are not an end in them, but serve as a means to achieve the business goals of a company. It is therefore essential for an evaluation or comparison of manufacturing concepts to identify the requirements on the manufacturing process against which the concepts should be evaluated. These requirements are derived from the business goals and the given or expected market conditions. Business goals and market conditions, however, may change over time and thus the set of manufacturing requirements may change. A manufacturing approach that has been sufficient until now may result in a poor performance in the future. Consequently, manufacturing concepts should be evaluated not only against the existing requirements, but also against future (possibly unknown) requirements. This section therefore looks at the current business trends and shows how these will change the manufacturing environment. The new manufacturing requirements are then used to derive requirements on the control of future manufacturing systems.

Recently, manufacturing industry has been facing a continuous change from a supplier's to a customer's market. The growing surplus of industrial capacity provides the customer with a greater choice, and increases the competition between suppliers. Aware of this power, the customer becomes more demanding and less loyal to a particular product brand. He/she demands constant product innovation, low-cost customisation and better service, and chooses the product which meets his requirements best.

In combination with globalisation, these trends will increase in the future. The consequences for the manufacturing industry are manifold. Companies must shorten product life cycles, reduce time-to-market, increase product variety and instantly satisfy demand, while maintaining quality and reducing investment costs.

These consequences imply:

- More complex products (because of more features and more variants),
- Faster changing products (because of reduced product life-cycles),
- Faster introduction of products (because of reduced time-to-market),
- A volatile output (in total volume and variant mix), and
- Reduced investment (per product).

The effects can be summarised as *increasing complexity* and *continual change* with *decreasing costs*.

3.13 APPLICATIONS OF AGENT TECHNOLOGY IN THE MANUFACTURING DOMAINS

3.13.1 Engineering Design

In the past decade, both the scale and the scope of engineering design had been changed and much enlarged. Design activities in various branches of engineering (mechanical, electrical, control, etc.) are now being integrated. Furthermore, acknowledging that engineering design must take into account the intrinsic requirements and properties of the processes that bring to life, create, maintain and re-cycle artifacts, concurrent engineering includes all the main life-cycle activities such as marketing, design, manufacturing, distribution, sales, operation, maintenance, and disposal and re-cycling. Participants in the product life-cycle can interact in parallel. Collaborative engineering (or design) transcends the above approaches: it emphasizes interaction instead of iteration, makes the conflicts between different stakeholders in the design process explicit and strives to achieve acceptable trade-offs via negotiation.

The above developments posed new requirements for the computational support of design. Decomposition and parallel execution in collaborative design, naturally, lend themselves to an agent-based approach. Beyond the usual advantages of having a modular system structure, additional merits are as follows: the agents need not be co-located, and they can form wrappers around and provide interface for existing legacy systems (e.g., analytical tools, simulators, CAD systems from various fields of engineering). Agents embodying knowledge and encapsulating tools of different engineering domains can communicate and work together if they have a set of shared concepts and terminology, a common language for expressing this knowledge, and a communication and control protocol for requesting information and services.

3.13.2 Process Planning

Process Planning is aimed at creating plans for discrete manufacturing operations that are executable in resource constrained production environments and produce the designed artifacts. Hence, Computer-Aided Process Planning (CAPP) incorporates both design and production related concepts: geometry, tolerances, surface quality, material properties, manufacturing processes, machines, tools and holding equipment (fixtures, grippers and robots). Useful domain knowledge varies also with the actual technologies such as machining, sheet metal bending, inspection, or (dis) assembly. There are two usual ways of handing the complexity of CAPP problems:

- De-structuring its world into manageable micro worlds – these are the so-called features.

- Decomposing the planning problem into sub problems such as process and resource selection, setup planning, sequencing, path planning, and NC programming.

In recent works, CAPP has been extended towards the execution of plans: planning is aware of available resources and takes actual lot-sizes or due dates into consideration.

Process planning –as a large body of planning problems in general – can be considered a problem of configuration. Configurational design and CAPP tend to apply similar representation and solution techniques that enable representing, solving and relaxing distributed constraint systems.

3.13.3 Production Planning and Resource Allocation

Production Planning is the process of selecting and sequencing activities so that they should achieve one or more goals of an enterprise and satisfy a set of domain constraints. At the strategic and long-term level, top managers try to allocate available resources based on their experience, intuition and computer support, if available. The project manager agent maintains the project milestones, the project activity network and each task's resource allocation information. A task agent is in charge of its own single task. A resource manager agent is in charge of monitoring and coordinating a set of resources. A coordinator agent is responsible for coordinating multiple resource allocation markets in the virtual market model.

3.13.4 Production Scheduling and Control

Scheduling is the process of selecting from among alternative plans and assigning resources and times to the activities in the plan. These assignments must obey a set of rules or constraints that reflect the temporal relationships between activities, the production technology and, the capacity limitations of shared resources. Manufacturing control relates to strategies and algorithms for operating a manufacturing plant, taking both the present and past observed states of the manufacturing plant, as well as the demand from the market into account. The manufacturing control problem can be considered at two levels. At the low-level, the individual manufacturing resources are to be controlled to perform actual processes expected by the high-level control functions. High level manufacturing control is concerned with coordinating the available manufacturing resources, in order to make the products required. In agent-based manufacturing systems, agent technology is usually applied to high-level manufacturing control, but can also be applied in the lower level.

Location of goods can be selected dynamically in near real-time. The higher level optimizer agent, with a global perspective, generates a balanced and synchronized order-tray sequence and efficiently assigns resources to each order tray, using a genetic algorithm (GA). The middle-level guide agent takes the resource assignment decision from the higher level agent and guides the lower level agents to achieve improved system performance. The lower level agents make their decisions based on real-time conditions, and suggest the alteration of predetermined resource assignments but have to obtain permission from the middle-level agent.

3.13.5 Process Control, Monitoring and Diagnosis

Process control, monitoring and diagnosis are closely related, partly overlapping fields. Monitoring involves observing, recording, and processing signals, and detecting abnormal conditions of the controlled process. Diagnosis is the process of generating plausible hypotheses on the causes that led to the current (abnormal) state of a system.

Monitoring, diagnosis and prognosis apply both to physical processes (e.g., at the machine level) and business processes (e.g., material and work flows). Signals in the physical process correspond to process parameters such as force, vibration, temperature, pressure. For business processes, material movements (e.g., from one location to another one), process completion times, and other transactional data associated with information or material flow serve as the signals. The frequency and range of physical process signals are much higher. While automatic (feedback) control is quite common for physical processes, business processes will likely require human intervention. Despite these seemingly wide differences, the scientific principles and even the techniques underlying monitoring, diagnosis and prognosis are similar in both the physical and business settings.

3.13.6 Enterprise Organization and Integration

Enterprise Integration is aimed at providing an Information Technology infrastructure for all business, engineering, operational and administrative functions of an enterprise that can be used for information exchange, decision making, coordination and collaboration. The integration efforts, usually, come together with organizational redesign and the re-engineering of business processes both within and between the main functional entities. Key ideas related to enterprise organization and integration is enterprise modeling, distributed planning and control, and information system modeling. Primarily due to globalization, the nature of businesses tends to be distributed making modeling, monitoring and control of business processes critical. Applying decentralized agents in enterprise integration permits the local parts of an enterprise to continue operation during temporary lapses in connectivity. Modeling formalisms include Petri nets, finite state machines, holons, and software agents. The most important questions here are:

- How to achieve the appropriate representation of process models,
- How to model various constraints within and between business functions such as marketing, design, planning, manufacturing,...

and material supply, and how to use them in order to find the best-of-practice process, and

- How to maintain interdependencies within the network of organizational entities.

3.13.7 Production in Networks

A Production (or supply) network is a net of suppliers, factories, warehouses, distribution centers and retailers through which raw materials are acquired, transformed and delivered to customers. In a supply network, the traditional boundaries of firms are dissolved: decisions on the use of resources should concern both internal and external capacities, and the internal flow of materials should be synchronized with the incoming and outgoing flows. There exist a number of Supply Chain Management (SCM) systems for integrating data of all major business functions at the nodes of a supply network, but these systems are rather transactional: they provide technology for information storing, retrieval and sharing, but do not really support decision making.

3.13.8 Assembly and Life-Cycle Management

Assembly, usually, represents the last technical step in the product creation process; however, according to the up-to-date organization principles, the ready-made products are to be followed throughout their life cycles. Hierarchical, heterarchical and holonic control structures for an assembly cell are compared in. Holonic systems were found to deliver better performance in a wider range of situations than their more conventional counterparts. For instance, the holonic concept demonstrated improvements in the robustness and volume flexibility in an engine assembly system in the automotive industry.

Agent technology and multi-agent systems have become prevalent in the past decade, enabled by a wide spectrum of information and control technology (such as networking, software engineering, distributed and concurrent systems, mobile technology, electronics commerce, interfaces, semantic web). Various agent technologies are attractive in all main domains of manufacturing because they offer help in realizing important properties as autonomy, responsiveness, modularity and openness. Multi-agent systems working in a decentralized way are able to use distributed and incomplete sources of information and knowledge. Still, uncertainties and eventual conflicts can be resolved via communication, collaboration and cooperation.

3.14 BENEFITS OF AGENT BASED APPROACH FOR MANUFACTURING

Though the agent-based approach allows for an open ended design and implementation of complex systems, the problems themselves cannot be solved by less effort, and scalability, safety and traditional software quality are serious bottlenecks. We have summarized the main barriers for the industrial take-up of agent technologies, such as the risk of consistent global operation, the appearance of inevitable conflicts between self-interested entities, and the extra burden of communication. Until recently, the industrial acceptance of MAS in manufacturing has been relative low, partly because of the above issues, and partly because of the difficulties in their stepwise integration with existing legacy systems. Developments in various agent technologies are still extremely dynamic, innovative and ramifying. The further evolution of multi-agent systems and manufacturing will probably proceed hand in hand: the former can receive real challenges from the latter, which, in turn, will have more and more benefits in applying agent technologies, presumably together with well-established or emerging approaches of other disciplines.

Agent-based manufacturing is a new way of thinking about and applying information. The primary benefits of the agent based approach are that they provide dynamic, reliable, and agile systems. As such, it will enable organizations of the future to accommodate rapidly changing business conditions; increase market responsiveness, lower cycle times, increase productivity, and better utilize its resources. In other words, the agent-based approach will be the way modern manufacturers develop their systems to compete in the twenty-first century.

Chapter 4

Agents for MRP, Process Planning and Scheduling

As outlined in the first chapter, while applying the agent based approach for supply chain management, for the purpose of effective information sharing agents are grouped into different modules, each group of agents in a module perform a major function of the enterprise. In this chapter the modules developed for MRP, process palming and scheduling activities have been highlighted.

4.1 FUNDAMENTALS OF MRP FOR RECOGNITION OF NECESSARY AGENTS

Material Requirements Planning is a time phased priority-planning technique that calculates material requirements and schedules supply to meet demands across all products and parts in one or more plants. Information Technology plays a major role in designing and implementing Material Requirements Planning systems and processes as it provides information about manufacturing needs as well as inventory levels. MRP techniques focus on optimizing inventory. MRP techniques are used to explode bills of material, to calculate net material requirements and plan future production.

A MRP systems use four pieces of information as shown in Fig. 4.1 to determine what material should be ordered and when

- The master production schedule, which describes when each product is scheduled to be manufactured.
- Bill of materials, which lists exactly the parts or materials required to make each product.
- Production cycle times and material needs at each stage of the production cycle time; supplier lead times.
- Inventory status file.

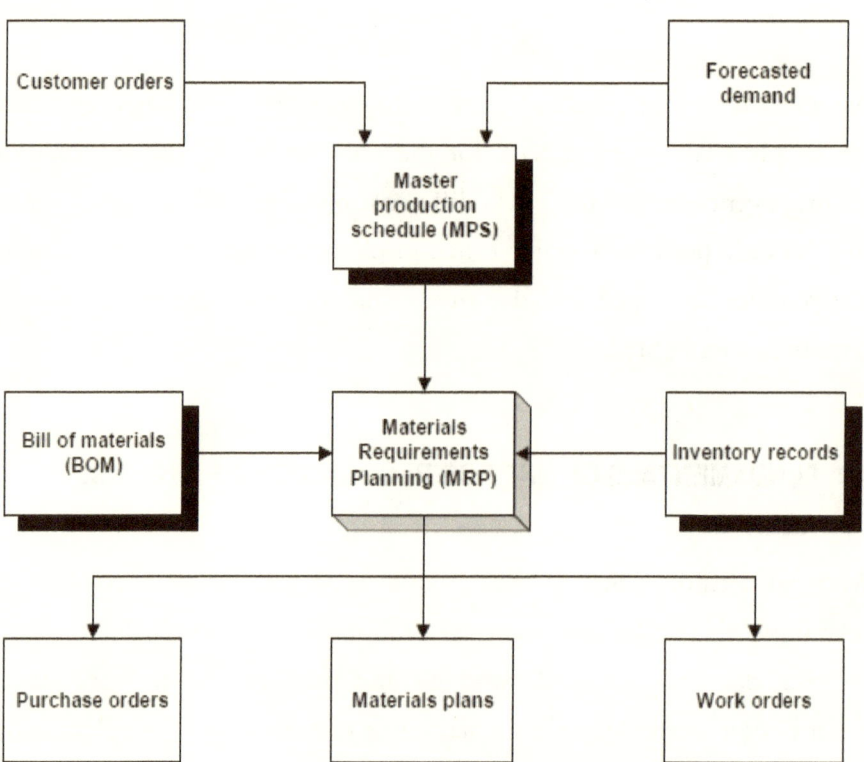

Fig. 4.1: Schematic View of Material Requirement Planning

The master schedule and bill of materials indicate what materials should be ordered. The master schedule, production cycle times and supplier lead times then jointly determine when orders should be placed.

The MRP has evolved to become a component of a MRPII system. Technically, MRPII extends MRP and links it with the company's information resources such as human resource information system, financial management, accounting, sales, etc. Such extension is typical according to modern trends in business management and modeling and made possible by advances in information technology. MRP systems lay in-between management control and operational control processes. However, as detailed production data are linked with overall organizational information resources it becomes clear that MRP and MRPII system implementations play a significant role in company's corporate advantage.

4.1.1 Inputs to MRP

4.1.1.1 *Master Production Schedule*

The Master Production Schedule (MPS) is the significant input that drives the MRP system. Primarily, MPS identify the quantity of the particular products that manufacturer is going to produce at a given time period. To achieve this, MPS needs combining two independent demands namely customer orders and forecasted demand. The MPS is also considered from other issues such as key capacity constraints; inventory levels and safety stock requirement (refer Fig. 4.2). MPS have to ensure that raw materials are available to meet the demands and MPS must not exceed the production plan or capacity plan. Turning to time frame for MPS, typically, a time bucket for MPS is one week. In addition, the minimum length of planning horizon should be equal to the longest lead time of item in process.

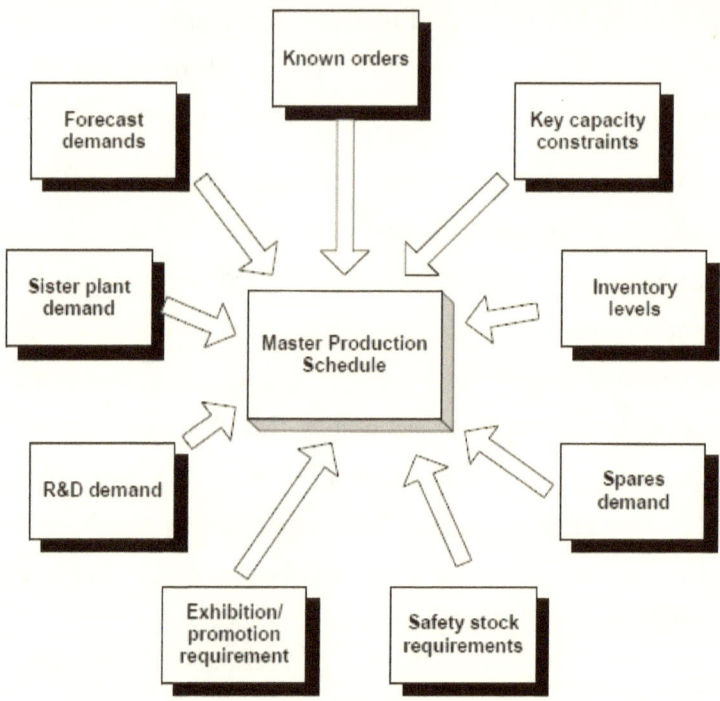

Fig. 4.2: Constituents for Master Production Schedule

4.1.1.2 *Bill of Material*

Bill of Materials gives information about the product structure, i.e., parts and raw material units necessary to manufacture one unit of the product of interest. Bill of Material (BOM) is another input of MRP system, which clarifies the structure of an independent demand item, a bill of material is a listing of all of the subassemblies, intermediates, parts, and raw materials that go into a parent assembly showing the quantity of each required to make an assembly. Bill of material facilitates MRP application checking the components for each item that is going to be produced.

4.1.1.3 *Inventory Data*

Inventory data facilitate MRP system identifying inventory status so as to calculate a 'net' requirement.

4.1.2 Conditions for Implementation

Several requirements have to be met, in order to given an MRP implementation project a chance of success. They are:

1. Availability of a computer based manufacturing system is a must. Although it is possible to obtain material requirements plan manually, it would be impossible to keep it up to date because of the highly dynamic nature of manufacturing environments.
2. A feasible master production schedule must be drawn up,
3. The bills of material should be accurate. It is essential to update them promptly to reflect any engineering changes brought to the product. If a component part is omitted from the bill of material it will never be ordered by the system.
4. Inventory records should be a precise representation of reality,
5. Lead times for all inventory items should be known and given to the MRP system.
6. Shop floor discipline is necessary to ensure that orders are processed in conformity with the established priorities.

4.2 AGENT BASED MRP

In the current section, all those agents which are responsible for the execution of activities related to MRP (Material requirements Planning) are described (refer Fig. 4.3). The following are the details of agents developed for this purpose.

4.2.1 Storage Agent

This agent is responsible for collecting the information regarding the available stock/components/material and to give access to the data to other agents through negotiation or by sending standard massages. The SDM consists of agents which are responsible for updating the data before and after each purchasing activity as well as supplying the required quantities of available materials to various departments. The updated information is available as output of this module and can be accessed by other agents Storage agent makes this information available to agents for time phasing.

4.2.2 Agent for the Generation of Master Production Schedule (MPSA)

Agent for Manufacturing Production scheduling play an important role for the MRP module and the whole process of time phasing process initiates from this stage for atomization. This agent basically starts receiving the information about customer orders, forecast data, available capacity and constraints for a given product from the concerned modules and agents namely CRM, MSM, and RCPA (refer Fig. 4.3). Thus received information is used for the generation of Master production Schedule for given product. This Agent transfers its output to TPA agent.

4.2.3 Agent for Rough Cut Capacity Planning (RCPA)

This Agent provides a Rough Cut capacity Plan, which will be developed after collecting the relevant data about the recourses from the related agents. This Agent also exploits the available capacity and prepares the capacity planning information required for agents of production planning activities in particular, scheduling.

4.2.4 Product Data Agent (PDA)

This agent is responsible for product data acquisition form PDDM regarding product structure, dimensions, tolerances and other specifications. This Agent is the basic Agent meant for collecting and storing almost all the required information about the product for any reference before, during, and after manufacturing activities. The Agent updates the information

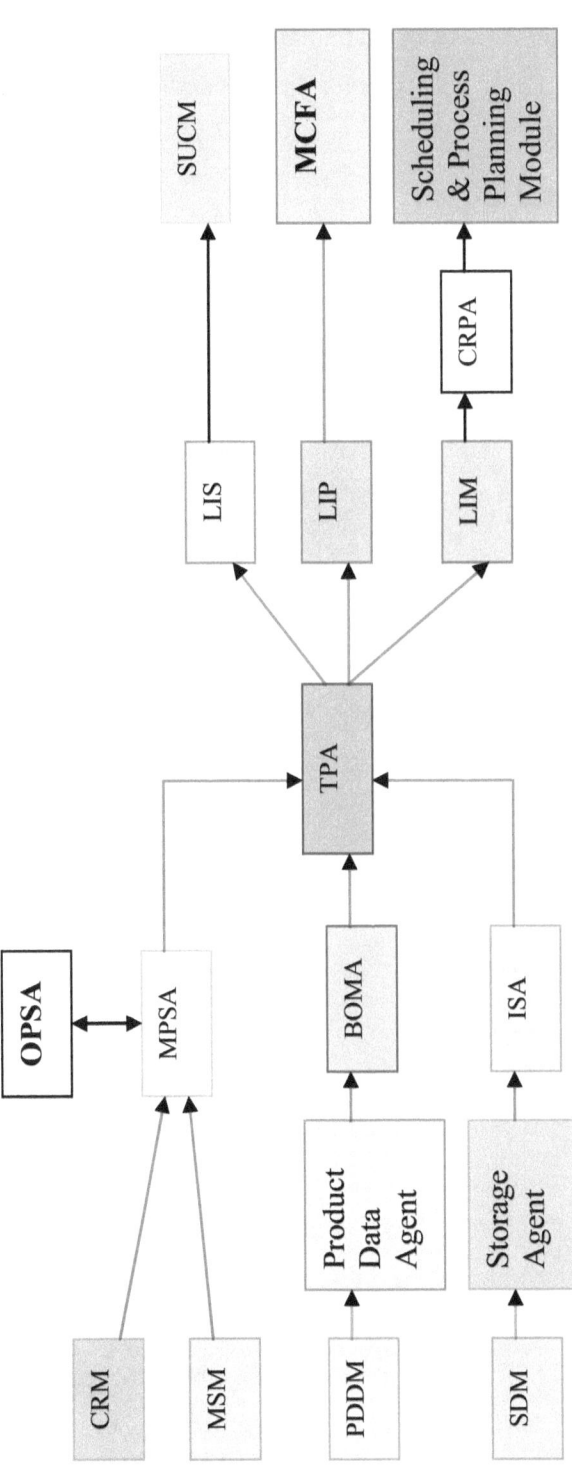

Fig. 4.3: Agents for MRP Activities

CRM- Customer requirement Module, MSM-Market Survey Module, PDDM-Product design & Development Module, SDM-Storage Data Module, RCPA-Rough cut Capacity Planning Agent, MPSA-Master Production Schedule Agent, BOMA-Bill Of Material Agent, ISA-Inventory status file Agent, TPA-Time Phasing Agent, LIS-List of Items for Subcontracting, LIP-List of Items for Purchasing, LIM-List of Items for Manufacturing, CRP-Capacity Requirement Planning Agent, MCFA – Machine Tool, Cutting Tool and Fixture selection Agent, SUCM-Subcontracting Module.

continuously so that the other Agents, in particular BOMA which need the product data for time phasing can access by sending predefined messages.

4.2.5 Agent for Bill of Material (BOMA)

This Agent obtains the necessary information about a given part from PDA and after processing this information computes the bill of material of the part. The bill of material clearly classifies the items as items to be purchased or manufactured or subcontracted. The Agent also does listing of assemblies and subassemblies parts and raw materials needed for a given end product.

4.2.6 Agent for Determination of Inventory Status files (ISA)

Inventory status file is one of the important inputs for time phasing of material requirements performed in MRP. It defines the available/levels of necessary inventory for the given part. ISA interacts with storage agent through standard message and collects the requisite information, then compiles it in the required format and presents it to TPA.

4.2.7 Agent for Time Phasing (TPA)

The functions of this agent includes collection of information for its tasks from other agents related to bills of material, product data, stores, master production schedule etc. This Agent has collection of logical procedures in order to processes the information through time phasing of planned orders for the determination of three lists, period wise list of items to be purchased (LIP), period wise items to be subcontracted (LIS) and period wise list of items to be manufactured (LIM). LIP will be sent to the agents of respective departments as well PAM to facilitate purchasing process. Similarly LIM will be submitted as input for PSM for scheduling of various items to be manufactured. LIS will be used by SUCM for its activities.

The logic employed by the TPA has been summarized in the flow chart and is displayed as Fig. 4.4.

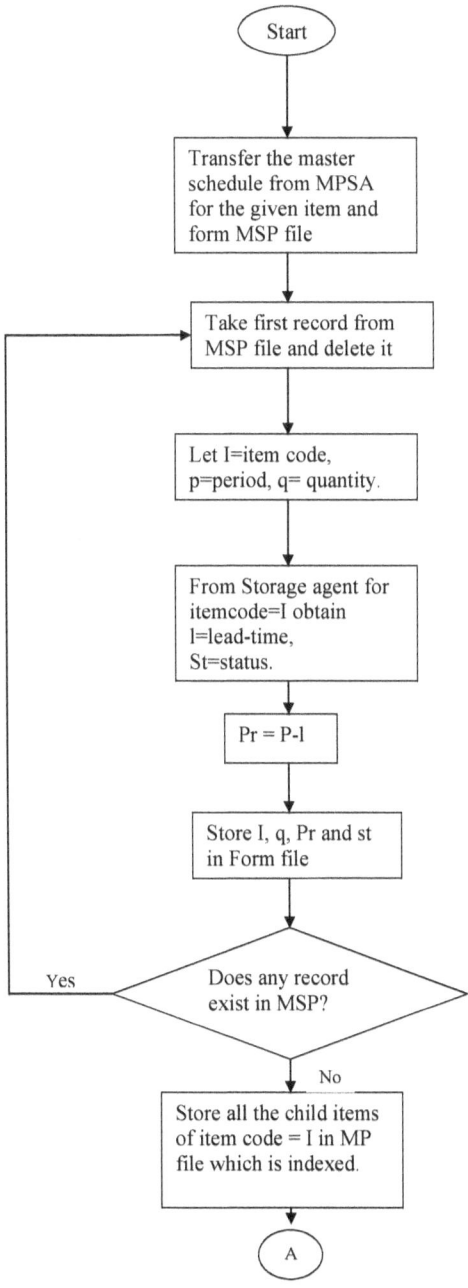

Fig. 4.4: Logic for Time Phasing

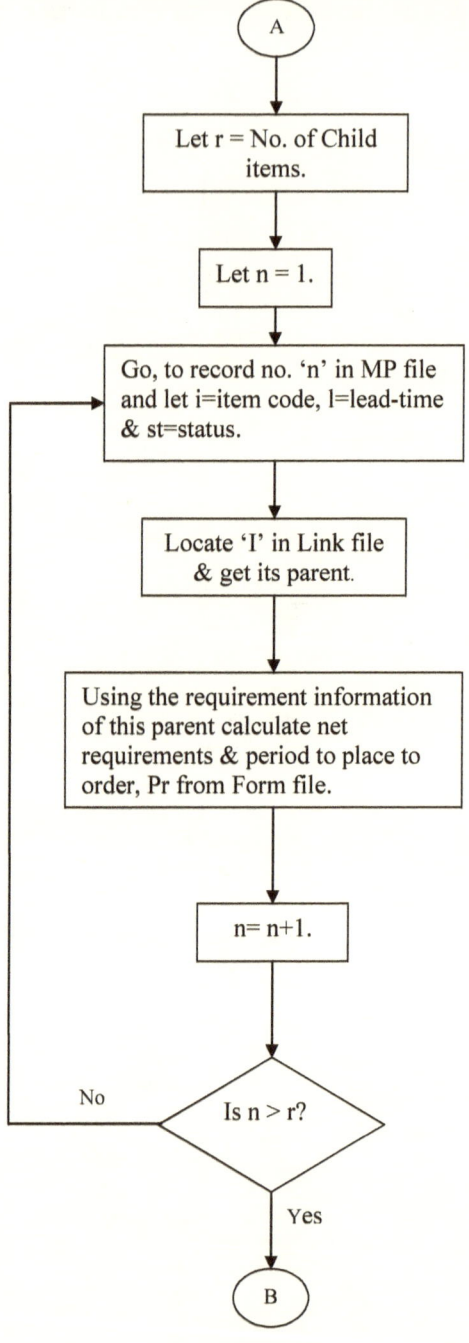

Fig. 4.4: (Continued)

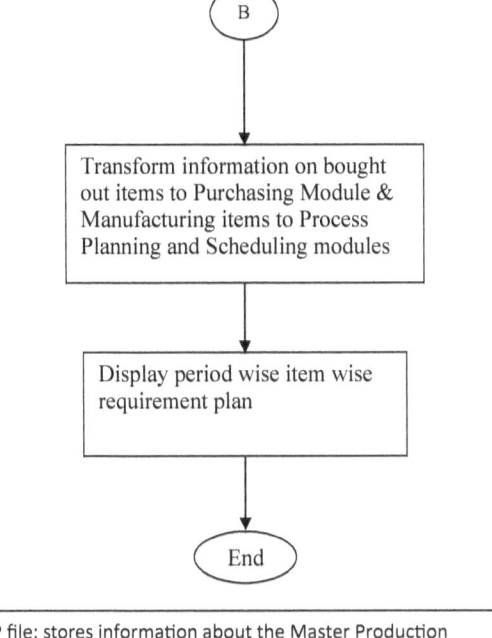

Fig. 4.4: (Continued)

As mentioned earlier, agents communicate and negotiate with the other agents, perform the operations based on the local available information and may pursue their local goals. The agents described in the previous section are expected to perform activities related to time phasing activities based on MRP concepts. The activities performed are development of rough cut capacity planning, master production schedule, time phasing of orders for the preparation of period wise list of items to be purchased, manufactured and subcontracted which lead to preparation of purchase orders as well as shop floor management (refer Fig. 4.3). This output has been made possible through effective sharing of information between all the agents.

4.3 FUNDAMENTALS OF PROCESS PLANNING

Process planning is a task aimed at translating part design specifications from an engineering drawing into manufacturing operation instructions. Conventionally, this task is carried out manually by an experienced process planner and the output is in the form of planning sheets. The process planner studies the engineering drawing in respect of features, dimensions, tolerances, surface finish, material and production volume etc. and then examines the available processes, machine tools, cutting tools, fixtures, etc. in order to decide the following: manufacturing processes, machine(s), and tool, machining parameters, fixture and sequence of operations. The quality of process plan influences directly the degree of complexity of the preparatory work for manufacturing good quality products, the production rates and costs and the degree of complexity of production scheduling. Hence process planning is of key importance to manufacturing.

4.3.1 Steps in Process Planning

A process engineer must follow the following well defined general steps:

(a) Preliminary analysis of the design.

(b) Selection of the initial blank and its manufacturing method.

(c) Selection of manufacturing processes for producing the given part.

(d) Selection of machining equipment and tooling for the operations.

(e) Determination of work piece setting method for each operation.

(f) Determination of operational dimensions (appearing in the operational sheet) and tolerances for machining operations.

(g) Selection of machining conditions and determination of time standards of each operation.

(h) Determination of the sequence of operations.

Computer Aided Process Planning is the application of computers to assist the human process planner in the process planning function. Interest in the use of computers to generate process plans for manufacturing of machined parts is worldwide. The manual process planning approach draws heavily upon the experience of process planner. In all cases the activity is highly subjective, labor-intensive and tedious and requires personnel who are well trained and experienced in the manufacturing practices. Even for an expert the preparation of a good and efficient process plan is a time consuming job. For mass produced goods the high cost of manual planning can be regarded as acceptable, because the resulting unit planning cost is not high. However with the increasing consumer awareness and globalization of markets, the present day manufacturing is characterized by large product variety and frequent design changes i.e. low quantity, high variety, small batch production. Therefore mass production is becoming increasingly rare and manual process planning is no longer economical.

In its lowest form, CAPP reduces the time and effort required to prepare process plans and provides more consistent process plans. In its most advanced state, it provides automated interface between CAD and CAM, thereby facilitating integration of various functions within the manufacturing system.

4.3.2 Basic Information Required for CAPP

The planning of a machining process is based on the following information which serves as the initial data for process planning:

- Drawing and specification of the part to be produced,
- Drawing of the blank from which the part is made,
- Available equipment and tooling for production and
- Relevant engineering and economic standards and handbooks.

Part drawing and its technical specifications provide the essential basis for process planning. The information comes from a blue print or a CAD file and includes part configuration, technical specifications (necessary dimensions, shape tolerances and surface requirements etc.) and material (required heat treatment and hardness etc.).

The blank making process is normally identified by the designer and shown in the drawing. There are different kinds of blanks made by different processes such as castings, smith and die forgings, cermets, welded blanks and blanks cut from rolled stock etc. It is necessary to know the blank manufacturing process and the shape, size and accuracy of the blank. These factors determine the allowances of all part surfaces to be removed and the methods of locating and clamping the workpiece in the initial stage of machining of the given part. When it first enters the machine shop, the blank should have at least one qualifying surface from which it can be located in the first operation. Hence, it is essential that the blank making process provides the necessary surfaces which are to be used as datum surfaces for the initial operation.

Process planning may be engaged under two different conditions: planning for a new plant/workshop and planning for an existing plant/workshop. In the former case it is possible to directly select the most reasonable machine tools, according to the manufacturing needs. In the latter case the machine tools should be selected from among the existing ones available in the plant. Hence the data concerning the machine tools should be available for process planning. This data can be obtained from the catalogues of the existing machine tools. Similarly the data on tooling which characterizes the tooling facilities available for production should be included in the CAPP system. This data can be obtained from tooling handbooks in which data on the standard work holders and tools is given.

With the process plans of jobs as input, a scheduling task is to schedule the operations of all the jobs on machines while precedence relationships in the process plans are satisfied. Although there is a close relationship between process planning and scheduling, the integration of them is still a challenge in both research and applications. Traditionally, process planning and scheduling were performed sequentially, where scheduling was done after process plans had been generated. Considering the fact that the two functions are usually complementary, it is necessary to integrate them more tightly so that the performance of a manufacturing system can be improved greatly.

4.4 AGENT BASED PROCESS PLANNING

The process planning procedure must coordinate the process planning functions, such as process selection, tool selection, feature sequencing, and machine tool selection without human intervention. These functions share some information, such as features of the part and machine tool parameters, and have distinct knowledge bases.

The complexity of a process planning process and the vast amount of information that a system would need to process to make planning decisions qualify process planning an appropriate candidate for being designed as a multi-agent planning system. For each of the above activities software based Agents are developed and can be used efficiently and effectively. Working in an integrated manner all the agents support in the automatic generation of efficient process plan for manufacturing a given component. The Fig. 4.5 shows the set of agents used in the generation of process plan.

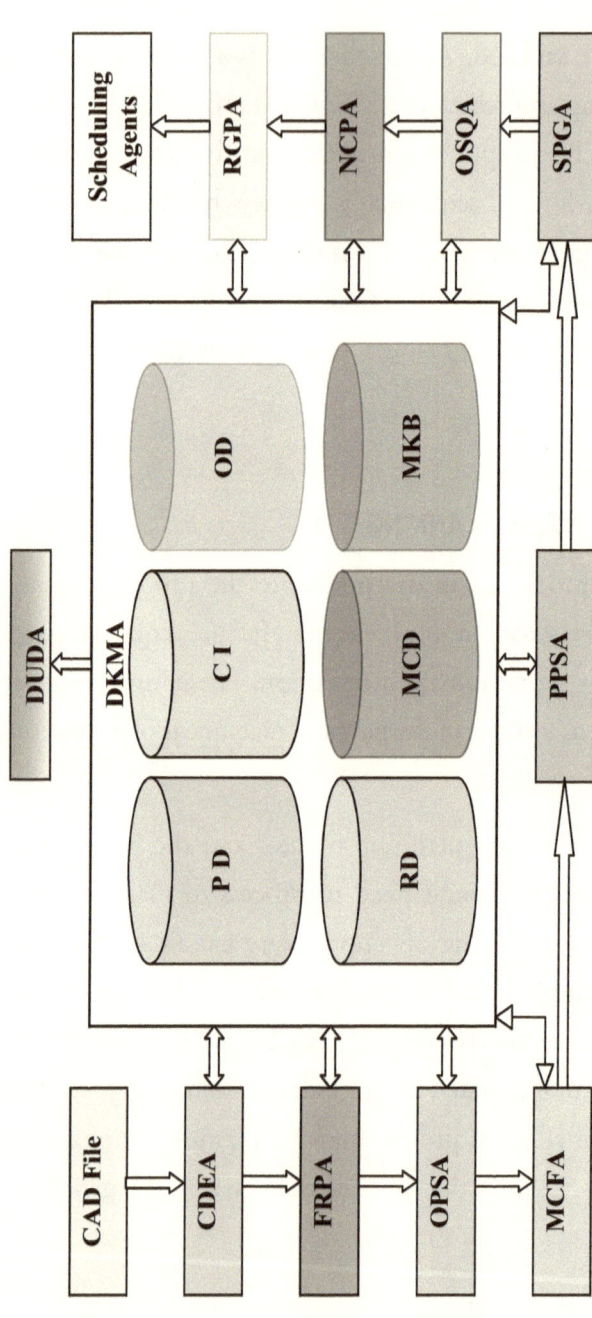

Fig. 4.5: Agents Identified for the Generation of Process Plan

Data Updating Display Agent(DUDA), Data and Knowledge base Management Agent (DKMA), Product Database (PD), Coded Information (CI), Operational Database (OD), Resource Database (RD), Machining Database (MCD), Machining Knowledgebase (MKB), CAD Data Extraction Agent (CDEA), Feature Recognition and Presentation Agent (FRPA), Operation Selection Agent(OPSA), Machine Tool, Cutting Tool and Fixture Selection Agent (MCFA), Process Parameter Selection Agent (PPSA), Setup plan Generation Agent (SPGA), Operations Sequencing Agent (OSQA), NC Program Generating Agent(NCPA)., Report Generation and Presentation Agent (RGPA).

4.4.1 Data and Knowledge base Management Agent (DKMA)

This agent is responsible for collection, storage and also updating of large amount of information in the following databases/knowledge bases that are necessary for the generation of process plan for a given part.

4.4.1.1 *Product Database (PD)*

This database stores all the design related and technological data about the component to be manufactured such as dimensional and geometric tolerances, topological information about the surfaces of the part etc.

4.4.1.2 *Coded Information (CI)*

This database has coded information in the form of predefined patterns and specific codes which are used to assign proper codes to manufacturing features, machine tools, cutting tools and fixtures. Such type of coding is necessary for the automation of process planning activities.

4.4.1.3 *Operational Database (OD)*

In this database detailed machining information, related to various types of machining features is organized systematically. For different machinable features of prismatic parts all the possible operations are identified along with the name of the machine tool and cutting tool required to perform that operation. For certain operation in involving complex features fixtures are also specified

4.4.1.4 *Resource Database (RD)*

This database is created to store information on available resources such as machine tools and cutting tools, fixtures and blanks of different sizes. The information has been compiled in coded form which makes it possible to carry out activities of selection of various recourses for machining operations automatically. This is consistent with today's manufacturing ethos where technology is changing at a rapid pace due to which facts and rules in the database require constant updating.

4.4.1.5 *Machining Database (MCD)*

The available data about the process parameters from various sources such as Handbooks and tool manufacturer's catalogues is coded and stored in the Machining Database. This database is used to obtain the recommended range of depth of cut, cutting speed and feed rates for a specified operation.

4.4.1.6 *Machining Knowledgebase (MKB)*

Machinists and process planners think and express themselves, mainly with rules of thumbs and heuristics, rather than precise equations and algorithms. The knowledge embodied in the rules of thumbs practiced by the machinists is well organized and stores in this database. From process knowledge stored in databases providing machining specific information, the process planning system can readily handle the diversity of tasks faced during generation of process plan.

For DKMA, it is necessary to input large amount of information as listed above related component to be machined, available resources, machining knowledge and machinability data. This agent permits the user to interact with all the data bases to perform the following functions:

(i) Updating the data bases: it is the function of adding fresh data or modifying existing data related to operations, resources, machining knowledge etc.

(ii) Deletion: it is the removal of some portion of data which has become irrelevant.

(iii) Retrieval: it is concerned with displaying any portion of the available data either for some verification or for providing an idea about the existing data.

4.4.2 CAD Data Extraction Agent (CDEA)

From the viewpoint of manufacturing, the biggest limitation of CAD systems is that the information stored in a CAD model is in the form of the modeling primitives: faces, edges and vertices which have been widely interpreted as low level information elements and can not be directly used for process planning. Manufacturing features such as slots, steps, holes, and pockets are generally regarded as high level information elements that are directly relevant in automated process planning. Hence, it is necessary to first develop a method of retrieving useful description of the part in terms of manufacturing features from a given CAD model.

This Agent is responsible for reading the CAD file of the given component in textual form and compile the data related to design of the part in a predefined form that support the activities of all other agents

4.4.3 Feature Recognition and Presentation Agent (FRPA)

For a product the manufacturing features are the volumes which are removed by one or a series of operations. They can therefore be called 'manufacturing features'. Design and manufacturing features in many cases differ from each other. Features provide natural means to associate domain knowledge i.e., given a part representation in terms of manufacturing features, process planning becomes essentially a discrete search problem because machining features are readily associated with manufacturing processes. Hence interpreting a work piece as a set of manufacturing features facilitates in the development of an automated process planning system.

Accordingly, feature recognition is the process of extracting manufacturing features from a CAD database and recognition of features is done using the geometric and topological information of CAD data by searching for the presence of a particular type of feature. For this FRPA interacts with CDEA to receive the CAD data of the component compiled in a form that support the recognition of manufacturing feature of the given part by using hint and rule based feature recognition method. To support the methodology a large number of rules are stored in computer compatible form in data base for coded information. This agent also performs the process of representing each recognized features in terms of certain characteristics, necessary for machining and setup planning activities.

4.4.4 Operation Selection Agent (OPSA)

Selection of the machining operation for each manufacturing feature is one of the important tasks in process planning. It is clearly related to the quality of the manufactured parts and the production efficiency and cost. In operation planning for the selection of operations, the requirements of a machining feature is mapped to the characteristics of a manufacturing operation.

In order to make a decision on the selection of operation, one must first understand the characteristics of the available operations. The basic characteristics of available operations generally include the shapes the operation can produce, the dimension limitations, surface finish attainable, geometric tolerance limitations and cost. Such characteristics are called the capability of a machining operation. Through their capability, the operations can be related to the requirements of a manufacturing feature. For this, the feature characteristics such as shape, size, accuracy and surface requirements are the main factors considered while selecting operations for its machining. By matching the available capability of an operation with the requirements of a feature, the necessary operations for the feature can be identified.

After collecting list of manufacturing features of the given part form FRPA, this agent browses the related databases by communicating the DKMA in order to identify all the possible operations.

4.4.5 Machine Tool, Cutting Tool and Fixture Selection Agent (MCFA)

The performance of any machine tool is bound by certain characteristics such as power, range of workpiece size that can be accommodated, process capability etc. which decide what types of surfaces a machine tool can produce. These characteristics not only specify or describe a machine tool, but also play an important role in determining its suitability for performing an operation required to produce a given feature. Therefore, the choice of the most suitable machine tool for an operation is a very important task as it greatly influences the machining accuracy, production rate and unit cost of the operation.

Similarly, cutting tools are central to the manufacturing process and tool selection is one of the most significant factors in deciding how a machined product will be manufactured. It has a direct impact on machinability and machine tool performance. In conventional process planning the tools are selected manually by an experienced user bearing in mind the various parameters which influence the machining process. The choice of the tool is subjective and often not the best. This factor together with increasing use of computers in manufacturing activities has highlighted the need for automatic methods of tool selection

For a given manufacturing feature, it is often possible to use different machine tools involving different operations. Further for each machine tool, several cutting tools suitable for machining a given feature may be available in a wide range of sizes. Many such combinations of machine tool and cutting tool provide a set of machining plans which meet the requirements of process capability, surface finish, production rate and cost to varying degrees of satisfaction. Hence selection of the right machine tool with appropriate cutting tool for each feature is one of the major decisions for the development process plan.

This Agent utilizes set of rules stored in machining knowledge base, interacts with DKMA to obtain the data about available recourses and finally identifies a suitable machine tool, cutting tool and fixture for a given operation.

4.4.6 Process Parameter Selection Agent (PPSA)

Selection of proper machining parameters (cutting speed, feed and depth of cut) for an operation is an important activity in machining planning as it affects productivity of an operation and cost because they determine MRR and also tool life. As MRR is increased machining cost is reduced. However, high MRR results in lower tool life and higher tool change cost. Tradeoff between MRR and tool life is needed for maximum productivity and minimum machining cost.

The common approach to achieve greatest efficiency of machining is to use the heaviest feed that will allow the required surface finish, use the maximum depth of cut consistent with power and then establish the cutting speed to give the desired life. Speeds may be increased to the point where the cutting edge will last only a few minutes or decreased to the point where the cutting edge will last for hours.

Three optimization criteria, generally considered for determining the value of speed are: i) minimization of cost, ii) maximization of the production rate and iii) maximization of the profit rate. Of these, the profit rate criteria require more information in terms of various costs which may not always be available to the process planning department. Hence the other two are more commonly used.

Hence, optimum selection of process parameters is critical. Generally an optimum set of parameters refers to the condition that combines long tool life and economic machining. However optimum conditions can differ from user to user based on his specific needs

This agent efficiently performs the process parameters selection activity for each operation by employing the combination of work material and tool material selected for the operation under consideration along with the name of the operation.

4.4.7 Setup Plan Generation Agent (SPGA)

A setup refers to a group of features that can be machined in a certain fixturing configuration. Setup planning is the process of grouping of features into setups, ensuring that all the features can be machined with specified accuracy. The combination of different set-ups that includes all the features of a given prismatic part is considered as a set-up plan.

This agent collects the necessary information form other agents and required data form DKMA then generates optimal setup plan using predefined methodology.

4.4.8 Operations Sequencing Agent (OSQA)

Operation sequencing is the function related to determining the order of performing a set of selected operations so that it satisfies the precedence constraints imposed both by the part and the operations. As a part contains many features, proper sequencing of operations selected for machining of these features is crucial in achieving efficient and high quality manufacture of the part. A machining sequence for a part should maximize product quality and minimize product cost under technological constraints immediately after SPGA generates optimal setup plan, these agents determines optimum sequence for operations.

4.4.9 NC Program Generating Agent

This Agent gathers information from all the agents and generates the NC program which will be used for processing of the given component.

4.4.10 Report Generation and Presentation Agent (RGPA)

This agent collects and compiles the details of process plan from the other agents and presents it in the standard format as optimal process plan for the machining the given component.

For executing its function this agent gathers three specific information form outputs of other agents. They are:

(i) List of recognized features, generated by the subsystem for feature recognition

(ii) Sets of feasible operations along with the information regarding machine tool, cutting tool and process parameters for each feature generated by subsystem for machining planning.

(iii) Optimal setups, order of executing the setups and sequence of operations within each setup generated by the subsystem for setup planning.

The consolidated output of above is the process plan for the given prismatic component, which correlates all the above mentioned outputs of other agents and then generates detailed process plan for machining the given component, clearly defining each manufacturing feature, specifying the recommended machine tool, cutting tool and process parameters for each feature and laying down the setup arrangements for machining all the features.

4.5 SCHEDULING

Scheduling is an important tool for manufacturing planning and it can have a major impact on the productivity of a manufacturing process. In manufacturing planning, the purpose of scheduling is to minimize the production time and costs, by allocating sufficient information to a production facility highlighting he details such as when to make, with which staff, and on which equipment. Production scheduling aims to maximize the efficiency of the operation and reduce costs.

Production scheduling tools provide the production scheduler with powerful graphical interfaces which can be used to visually optimize real-time work loads in various stages of production. Companies use both backward and forward scheduling to allocate plant and machinery resources, plan human resources, plan production processes and purchase materials. Forward scheduling is planning the tasks from the date resources become available to determine the shipping date or the due date. Backward scheduling is planning the tasks from the due date or required-by date to determine the start date and/or any changes in capacity required.

Agent-based scheduling system addresses many of the limitations of existing job-shop scheduling systems by providing a decision-support framework for scheduling that naturally accounts for conflicting and changing goals. By providing mechanisms for manual intervention and evaluating the impacts of such changes, the scheduler can react sensibly to near-term uncertainties and contribute expert knowledge to the system. In this way, the human scheduler essentially acts like one of the software agents, participating on an equal basis in the formulation of candidate alternatives.

A feasible optimal schedule can be generated according to the algorithms and logical functions of the scheduling agents, so as to minimize the total resource consumption with make span constraints in scheduling. And the proposed agent-based approach with learning effect can provide another perspective domain and concept for solving scheduling problems.

4.5.1 Scheduling Algorithm

For scheduling module algorithm proposed by Giffler and Thompson has been used. It can be explained as follows:

Let,

PS_t = a partial schedule containing t schedule (operations)

S_t = the set of schedulable operations at stage t corresponding to a given PS_t.

σ_j = The earliest time at which operation J S_t could be started.

$\emptyset_j$ = the earliest time at which operation J S_t could be completed.

For a given active partial schedule the potential start time σ_j is determined by the completion time of the direct predecessor of operation j and the latest completion time on the machine required by operation j. the larger of these two quantities is σ_j. The potential finish time $\emptyset_j$ is $\sigma_j + t_j$ where t_j is processing time of operation j.

The procedure starts as follows:

Step 1: Let t = 0 and begin with PS_t as the null partial schedule. Initially S_t includes all operated with no predecessors.

Step 2: Determine $\emptyset^*$ min $_{jest}$ [$\emptyset_j$] and the machine m* on which * could be realized.

Step 3: For each operation J ε S_t that require machine m* and for each $\sigma_j < \emptyset^*$ create a new partial schedule in which operation j is added to PS_t are started at time σ_j.

Step 4: For each new partial schedule PS_{t+1} created in step 3 update the data set as follows:

(a) Remove operation j from S_t.

(b) Form S_{t+1} by adding the direct successor of operation j to S_t.

(c) Increment t by one.

Step 5: Return to step 2 for each PS_{t+1} and continue in this manner until all active schedules have been generated.

4.5.2 Capacity Requirement Planning

After developing the schedules, next step is to check whether the available capacity is sufficient to meet the required capacity or not. If the required capacity is more than the available capacity and even rescheduling is not possible, then an analysis should be done to add new shift or overtime or add some more resources or subcontracting some non-critical jobs. Some times typical control policies are followed such as production switchover policy in which two levels of production high and low are specified. A control limit CL is also specified for number of hours in the work centre queue at the beginning of the period. If the number of standard hours in the work centre queue at the beginning of the week exceeds CL then high production level is implemented.

4.6 AGENTS FOR SCHEDULING

Agent-based scheduling module consists of a set of agents as shown in Fig. 4.6 which can interact with each other to share and transfer the required information and collectively execute all the activities that are necessary for generation of optimum schedule for performing a set of jobs in a given period of time. In principle, each agent can reach any of the other agents in the system by sending messages. The description each agent is given below.

4.6.1 Scheduling Agent (SCHA)

The SCHA performs the main scheduling function in the scheduling module and hence is treated as the important agent of this module which receives information about the assignment of jobs to be scheduled in a given period, obtains information about the sequence of jobs from SQVA and operation times form STCA.

It generates optimum schedule for the given set of jobs by employing the suitable algorithm for both flow shop and job shop cases. Foe job shop case it utilize he algorithm proposed by Giffler and Thompson (refer section 4.5.1). While developing optimal schedule its major stress will be on availability of jobs and availability of recourses. During this process it interacts continuously with ARIA for the information about the availability of resources.

It works with SEXA in implementing the optimal schedule generated by it and at the shop floor level it cooperatively works with SFCA in achieving distributed shop floor scheduling

4.6.2 Available Resource Information Agent (ARIA)

This agent collects information about assignment of jobs for a given period, and then interacts with the agents who are managing supporting databases to identify the manufacturing resources that are essential for performing the given jobs. First, it lists all the resources which satisfy both technical and availability requirements and then form this list it selects most suitable recourses and recommends this information to SCHA as well as CPRA

4.6.3 Capacity Requirement Planning Agent (CRPA)

This agent is responsible for the calculation of capacity required for executing a set of jobs in a given period. Than it interacts with ARIA to gather information about the availability of various resources in the period under consideration and make an estimation of available capacity. If the available capacity is less than the required capacity this agent indicate the alternative provisions such as rescheduling of the jobs by deleting some non-critical jobs (which may be subcontracted) or exploiting the possibilities of utilizing the available capacities (if available) either in the preceding period or in the succeeding period.

4.6.4 Sequence Verification Agent (SQVA)

Once a set of jobs to be scheduled is assigned to SCHA it communicates to this agent i.e. SQVA for verification of sequence in which the jobs under consideration are to be executed. If necessary this agent interacts with OSQA of process planning module for performing its activities. OSQA has a set of protocols and well define procedures for determining the optimal sequence for a given set of jobs.

4.6.5 Standard Time Calculation Agent (STCA)

This agent has the capability to determine the operation time required for any given operation. It makes use of standard data stored in predefined form which can be processed for the determination of operation time through standard procedures. After receiving a message and request form SCHA with a list of jobs, this agents calculates and provides the operation times for each of the given set of jobs.

4.6.6 Schedule Execution Agent (SEXA)

The main function of this agent is it accepts production schedules from the scheduler agent SCHA and distributes them to resources which are identified by ARIA. The optimum schedule generated by the SCHA will be transferred to this agent who assumes the responsibility of assigning and loading the jobs to the respective resources as per the schedule, for performing the jobs which are listed in the schedule.

4.6.7 Shop Floor Control Agent (SFCA)

SFCA is the agent that represents the overall control. SEXA report the working status of all the resources and the status information, including routine processing data and unexpected disruptions. It monitors the processing status of resources, analyzes and aggregates the processed data. If unexpected changes affect the execution of the schedule, it will report to the scheduler with high level scheduling related processing information to take appropriate action.

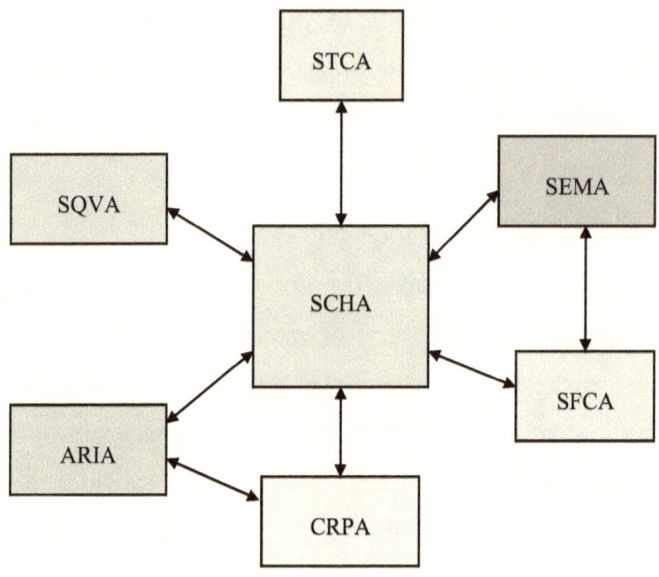

Fig. 4.6: Scheduling Agents

SQVA-Sequence Verification Agent,
SEXA-Schedule Execution Agent,
SCHA-Scheduling Agent,
ARIA-available resource Information Agent,
STCA-Standard Time Calculation Agent,
SFCA-Shop Floor Control Agent,
CRPA-Capacity Requirement Planning Agent,

Chapter 5

Multi Agent Framework for Purchasing Process

Purchasing is one of the basic functions of each company and it is essential for a company to work in its business environment. It is often said that the objective for the purchasing is "to acquire the right quality of material, at the right time, in the right quantity, from the right source, at the right price."

Before the description of various agents that have been developed for performing purchasing activities automatically, a brief discussion about the information contents used for the development of agents are given below.

5.1 STEPS IN PURCHASE PROCESS

There has been an evolution in the role and structure of the purchasing function through the nineties. The purchasing function has gained great importance in the supply chain management due to factors such as globalization, increased value added in supply, and accelerated technological change.

Purchasing involves buying the raw materials, supplies, and components for the organization. The activities associated with it include selecting and qualifying suppliers, rating supplier performance, negotiating contracts, comparing price, quality and service, sourcing goods and service, timing purchases, selling terms of sale, evaluating the value received, predicting price, service, and sometimes demand changes, specifying the form in which goods are to be received, etc.

A key and perhaps the most important process of the purchasing function is the efficient selection of suppliers, because it brings significant savings for the organization. The objective of the supplier selection process is to reduce risk and maximize the total value for the buyer, and it involves considering a series of strategic variables.

The following are the major steps in involved in executing the purchasing function in an organization which are arranged in sequence with explanations.

5.1.1 Requisition Generation & Approval

As part of an organization's internal financial controls, the accounting department may institute a purchase requisition process to help in managing requests for purchases. Requests for the purchase of goods and services are documented and routed for approval within the organization and then delivered to the accounting group. A purchase requisition is an authorization for a purchasing department to procure goods or services. It is originated and approved by the department requiring the goods or services. Typically, it contains a description and quantity of the goods or services to be purchased, a required delivery date, account number and the amount of money that the purchasing department is authorized to spend for the goods or services. Often, the names of suggested supply sources are also included.

A purchase requisition is owned by the originating department and should not be changed by the purchasing department without obtaining approval from the originating department. This important distinction is not clearly defined in some of the more popular integrated procurement software systems on the market today. In some industrial environments, the purchasing department may be assigned responsibility for requesting and purchasing goods. This is especially true for raw material purchases where the purchasing department is also responsible for inventory management.

5.1.2 Identification of Suppliers

Identifying potential suppliers is different from reaching a contract or agreement with suppliers. Once the need and the description of the need are identified, one of the following can happen

a) The need is fulfilled by a supplier that has an existing contractual relationship with the buying company.

b) The need is fulfilled by a new supplier that is not currently qualified to provide product and services to the firm.

In the first case purchasing personnel have already identified which suppliers will be used to source the need, and they have already taken steps to evaluate and prequalify the supplier. Qualification is important, as the purchasing firm must ascertain that the supplier meets several criteria and evaluate whether it is qualified to do the business and meet the needs of their internal customers in a satisfactory manners.

In second case where the supplier is not identified, or when the internal customer requests that the need be fulfilled by a specific supplier of their choice, purchasing face a more difficult challenge. Because there is no existing contract with the supplier, they may balk at approving the need fulfillment from this supplier. When internal customers purchase directly from nonqualified suppliers and try to bypass purchasing in the process, this is known as maverick spending. With little risk the maverick spending may be allowed up to certain limit.

For some items, firm may maintain a list of preferred suppliers that receive the first opportunity for new business. A preferred supplier has demonstrated its performance capabilities through previous purchase contracts and therefore receives preference during the supplier selection process, by maintaining preferred supplier list; purchasing personnel can quickly identify suppliers with proven performance capabilities. In cases when there is not a preferred supplier available, purchasing must get involved in selecting a supplier to fulfill that need.

5.1.3 Evaluation of Suppliers & Bidding/Negotiating

Supplier evaluation is a continual process within purchasing departments and forms part of the pre-qualification step within the purchasing process; although in many organizations it includes the participation and input of other departments and stakeholders. It often takes the form of either a questionnaire or interview, sometimes even a site visit, and includes appraisals of various aspects of the supplier's business including capacity, financials, quality assurance, organizational structure and processes and performance.

Based on the information obtained via the evaluation, a supplier is scored and either approved or not approved as one from whom to procure materials or services. In many organizations, there is an approved supplier list (ASL) to which a qualified supplier is then added. If rejected the supplier is generally not made available to the assessing company's procurement team. Once approved, a supplier may be reevaluated on a periodic, often annual, basis.

The easiest way to evaluate suppliers is to assess prior performance with existing suppliers. If the supplier to be evaluated is new, more investigation is required. The most common method is to find answer to following questions:

- How long has the supplier been in business?
- Who are the principal owners? If the company is privately held, are the owners also the current managers? If the supplier is a public corporation, obtain a copy of its current annual report.
- Who are the supplier's major customers? May you contact some of them for references?
- What have been the supplier's business trends over the past ten years?
- What is the history in labor relations? Is there a union? When is the contract due for renegotiation?
- What percent of sales is spent on research and development?
- What quality – control systems do they use?
    a) Reliance on inspection
    b) Process controls (SQC)
    c) Self-monitoring processes
    d) None
- What is the current backlog and delivery lead time?
- What is its history of price changes?

One of the important processes of evaluation of suppliers is through bidding or negotiating the supplier. This is the process where the organization identifies potential suppliers for specified supplies, services or equipment. These suppliers' credentials (qualifications) and history are analyzed, together with the products or services they offer. The bidder selection process varies from organization to organization, but can include running credit reports, interviewing management, testing products, and touring facilities. This process is not always done in order of importance, but rather in order of expense. Often purchasing managers research on potential bidders and obtain information on the organizations and products from media sources and their own industry contacts.

This is the process an organization utilizes to procure goods, services or equipment. Processes vary significantly from the stringent to the very informal. Large corporations and governmental entities are most likely to have stringent and formal processes. These processes can utilize specialized bid forms that require specific procedures and detail.

The very stringent procedures require bids to be open by several staff from various departments to ensure fairness and impartiality. Responses are usually very detailed. Bidders not responding exactly as specified and following the published procedures can be disqualified. Smaller private businesses are more likely to have less formal procedures. Bids can be in the form of an email to all of the bidders specifying products or services. Responses by bidders can be detailed or just the proposed dollar amount.

The bid usually involves a specific form the bidder fills out and must be returned by a specified deadline. Depending of the commodity being purchased and the organization the bid may specify a weighted evaluation criterion. Other bids would be evaluated at the discretion of purchasing or the end users. Some bids could be evaluated by a cross-functional committee. Other bids may be evaluated by the end user or the buyer in Purchasing.

Negotiating on the other hand is a key skill set in the Purchasing field. One of the goals of Purchasing Agents is to acquire goods as per the most advantageous terms of the buying entity (Buyer). Purchasing Agents typically attempt to decrease costs while meeting the Buyer's other requirements such as an on-time delivery, compliance to the commercial terms and conditions (including the warranty, the transfer of risk, assignment, auditing rights, confidentiality, remedies, etc).

5.1.4 Selection of Suppliers

The important selection process of suppliers is Vendor rating or Supplier rating. It is a business term used to describe the process of measuring an organization's supplier capabilities and performance. Supplier rating often forms part of an organization's supplier relationship management program. Such systems can vary in the criteria that are assessed. Both quantitative and qualitative types can be used for vendor rating; the process varies from one organization to another. Common criteria often include:

- Quality – for example number of not right first-time deliveries
- Delivery schedule adherence
- Cost/Price
- Capability
- Service

Results of each variable are then weighted into a final score – usually a percentage, allowing suppliers to be ranked. Various criteria can be analyzed within supplier rating systems – a common approach is to utilize

Cost Quality and Delivery measures and apply weighting against criteria in accordance with company requirements.

5.1.5 Purchasing Approval

Unless the company has self-approval purchase process, purchase requests for products or services must be approved by responsible person prior to Market Assessment. Financial approval is required for various reasons, for example, to prevent people from spending more money than allowed by the budget. Departments and divisions may have a variety of purchasing hierarchies and different responsible people depending on allowed limit from hundred up to several thousand Rupees or depending on the source of funds for the expenditure.

Due to various levels of approval required it takes so long to make a Purchase decision. Each company establishes its own procedure for obtaining approval. It determines the terms of approval to ensure that no unnecessary delay is caused to requesting purchase. If a purchase request does not receive approval, the procedure prescribes further actions, so that the Ordering Process could take place after certain changes in requisition form. Financial approval must be given before the purchasing commitment is made, and the purchasing system should be designed to ensure that this is done.

5.1.6 Purchase Order Release

A purchase order (PO) is a commercial document issued by a buyer to a seller, indicating types, quantities, and agreed prices for products or services the seller will provide to the buyer. Sending a PO to a supplier constitutes a legal offer to buy products or services. Acceptance of a PO by a seller usually forms a one-off contract between the buyer and seller. So the purchase order becomes an important entity in the process of completion of purchasing activity with it impact on other aspects and activities of the organization. Hence the release of purchase order has to have a strategy of optimization of economy and speed of response which is very important in Supply Chain management.

5.1.7 Expediting and Delivery from Supplier

Expediting is a concept in purchasing and project management for securing the quality and timely delivery of goods and components. The procurement department or an external expediter controls the progress of manufacturing at the supplier concerning quality, packing, conformity with standards and set timelines. Thus the expediter makes sure that the required goods arrive at the appointed date in the agreed quality at the agreed location.

To save the unnecessary costs, the supplier and customer may agree on the use of a third party expediter. These are specialists from companies specializing in this field who keep track of the deadlines and whether the components are properly packed. After inspection they notify the involved parties about their findings.

During the purchasing process one of the common tasks to be completed is the delivery schedule typically this will be outlined during the quotation phase and confirmed at time of order placement. This is usually in the form of an agreed lead time or a specified date/time.

The expediting process refers to the procurement organization contacting the supplier for either updates on the delivery schedule or to reassess the schedule based on issues with the supply of parts. The challenge for this process and one which is a common issue is who actually chases the supplier. In many cases this will typically be the buyer that raises the order – they may have a regular process where they reconfirm the order book or a specific order.

Where there are many stakeholders relying on delivery – in a manufacturing organization for example – there may be a temptation for everyone to get involved in this process where the part becomes critical – or is late enough to hold up production. Where problems continue and poor performance becomes an issue with a supplier then a suitable member of the management team should become involved.

5.1.8 Supplier Invoice Paid

An invoice or bill is a commercial document issued by a seller to the buyer, indicating the products, quantities, and agreed prices for products or services the seller has provided the buyer. An invoice indicates the buyer must pay the seller, according to the payment terms. The buyer has a maximum amount of days to pay these goods and are sometimes offered a discount if paid before.

From the point of view of a seller, an invoice is a sales invoice. From the point of view of a buyer, an invoice is a purchase invoice. The document indicates the buyer and seller, but the term invoice indicates money is owed or owing.

A typical invoice contains:

- The word invoice (or Tax Invoice).
- A unique reference number.
- Date of the invoice.
- Tax payments if relevant.
- Name and contact details of the seller.
- Tax or company registration details of seller (if relevant).
- Name and contact details of the buyer.
- Date that the product was sent or delivered.

- Purchase order number.
- Description of the product(s).
- Unit price(s) of the product(s) (if relevant).
- Total amount charged and
- Payment terms.

5.1.9 Update Supplier Information

For a large manufacturer, suppliers are key members of their team when it comes to fulfilling contract requirements and providing quality products and services to customers. Manufacturing organizations need to manage information about the suppliers of each component of their product as part of "supply chain management." Managing supplier directories, which contain contact information, repair, warranty and other overall support information, and making them available to both internal users and its customers is a critical activity for manufacturers.

Manufacturers need a supplier information system that will;

- Reduce the amount of time spent in tracking down and updating supplier information and also improve the quality of the supplier information.
- Enable the integration of the data set with other information products which may require the information, such as parts catalogs or user manuals.
- Improve the accessibility of the information for suppliers, customers and internal users.

Figure 5.1 present a flow chart describing purchasing process.

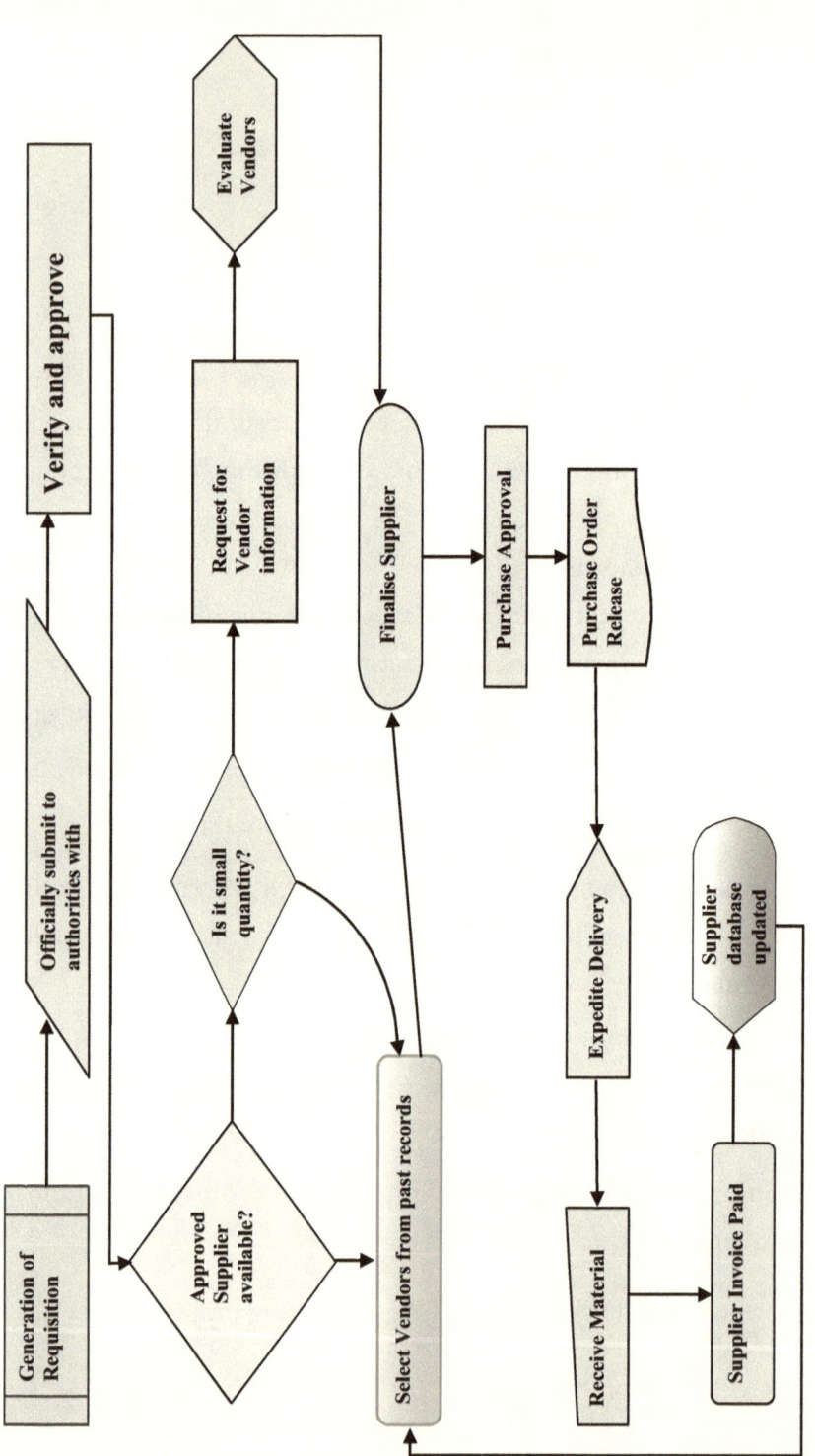

Fig. 5.1: Purchasing Process

5.2 PROPOSED AGENTS FOR THE AUTOMATION OF PURCHASING PROCESS

Purchasing module consists of 10 agents and each AGENT is designed to perform a specific function. Each AGENT uses common database and common form for collection and display of data as shown in Fig. 5.2. These agents have been described in the following sections

5.2.1 AGENT 1 – Availability Checking & Material Requisition Agent (ACMRA)

For the purpose of acquiring required material and supplies from each department an employee is authorized in each department to use the facility of logging in to view storage database for the verification of the materials available in store.

This agent permit the authorized user form all departments to check availability of any materials in the store and then order the items if sufficient quantity are available in store. If sufficient quantity is not available, user can generate and submit material requisition according to his requirements. Hence, the output form this agent is either Available Quantity (AQ) or Material Requirement form (MRF).

When the authorized users from any departments logs in to check availability of any materials in the store, the employee database is scanned to confirm the identity and permit the user to view available quantity. The database of employee includes fields like employee name, employee ID, department, and designation and employee Key number

Employee Key Number is connected with employee ID card through a combination of software and hardware. This Employee Key Number will be modified in every 15 days and informed to respective employee. The employee should enter the new updated key number, which is communicated to him.

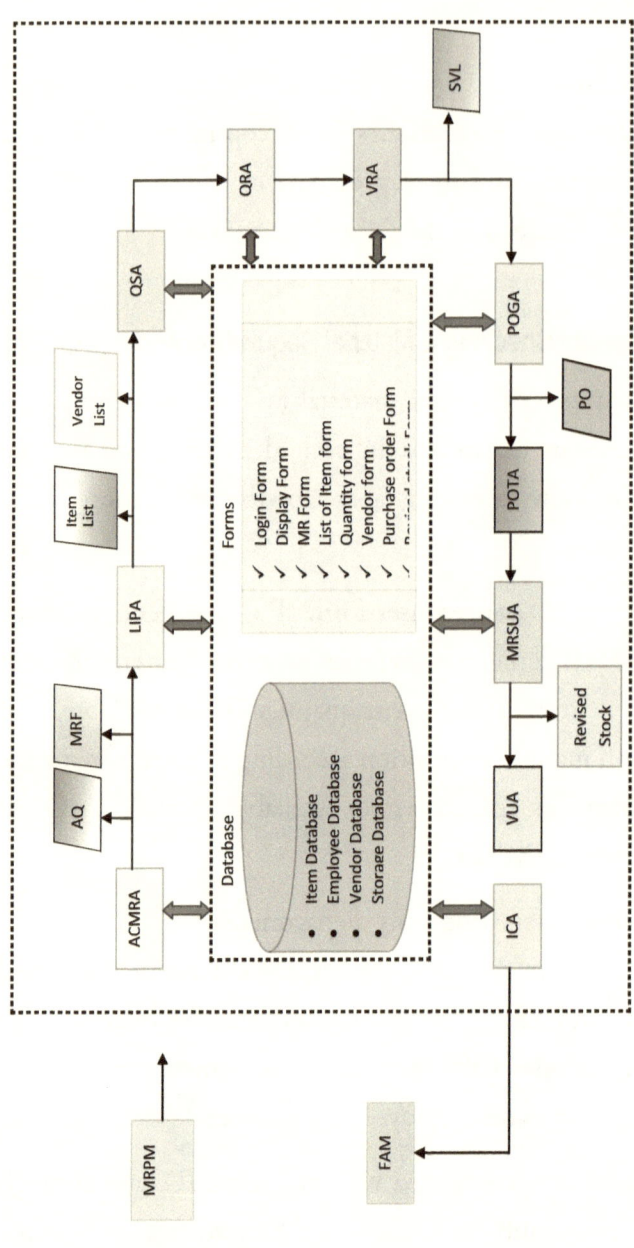

Fig. 5.2: Agents Identified for the Generation of Purchasing Process

ACMRA- Availability Checking & Material requisition Agent, LIPA- List of Items to Purchase Agent, QSA-Quotation Sending Agent, QRA- Quotation Receiving & listing Agent, VRA-Vendor Rating Agent, POGA-Purchase Order Generation Agent, POTA-Purchase Order Tracking Agent, MRSUA-Material Receipt & Stock Updating Agent, VUA-Vendor data Updating Agent, ICA-Invoice clearance Agent, AQ-Average Quantity, MRF- Material Requisition Form, SVL-Selected Vendor List, PO-Purchase Order.

Once, an employee successfully logs in to the application items available in the store will be displayed. The database used during this process is 'Item database' which consists of several fields shown in Table 5.1. The available quantity indicates the quantity existing in the store.

There may be some pending orders already requested by other users. Based on these pending orders the net quantity shows quantity that may remain in the store after satisfying the pending orders. This also becomes actual quantity available for the new user. The net quantity for the material under consideration will enable the user to take decision regarding ordering the quantity directly from the store or send material requisition form for purchasing of the items.

Table 5.1: Contents of Item Database

Item Code *Alphanumeric*	Item Name *Characters*	Available Quantity *Numeric*	Pending Requisitions *Numeric*
Net quantity *Numeric*	Average Lead Time *Numeric*	Average ordering Quantity *Numeric*	Average ordering Cost *Numeric*
Average Storage cost *Numeric*	Average Shortage cost *Numeric*	Average Annual usage *Numeric*	Previous vendor *Characters*
Preferred vendor *Characters*	Status for storage *Numeric*	Status for Priority *Numeric*	Dimensions *Numeric*
Measured units *Numeric*	Cost per unit *Numeric*	Major Characteristics *Alphanumeric*	End Products *Characters*

It means if net quantity is adequate as per his/her requirements the authorized user can choose the option of ordering form store to obtain the required quantity. "Items in store" form will be displayed after employee successfully log in to the application.

On the other hand if sufficient quantity is not available, user can generate and submit material requisition according to his requirement by choosing the related option.

5.2.2 AGENT 2 – List of Items to Purchase Agent (LIPA)

This agent receives message form ACMRA and collects all Material requisitions from various departments. It then makes separate lists of all common items ordered by different departments as well as items ordered by single department. Using suitable predefined rules for common items it determines total quantity and due dates.

For example let three departments D1, D4 and D9 have sent requisition for same material say M1. Calculation of common requirement and duration for this particular material M1 will be carried out by arranging the individual requirements as shown below in the Table 5.2.

The quantity to be purchased will be sum of all individual requirements and accordingly the ordering quantity will be determined by pre defined purchasing polices. The due date is calculated on the basis of earliest requirement among depending up on the individual due dates.

Table 5.2: Determination of Quantity and Due Date for Item Required for Multiple Departments

Dept	Quantity	Duration
D1	44	15
D4	16	16
D9	63	23
Total quantity & duration for m1	123	15

5.2.3 AGENT 3 – Quotation Sending Agent (QSA)

This agent is responsible for identifying vendors for each item to be purchased form the available vendor list (From Vendor database) or acquiring information about new vendors. The available vendor list also includes the names of all those vendors which are suggested by individual departments. Every vendor listed in the existing vendor database is assigned a grade based on the performance in supply of material in all the previous orders.

Any vendor having been allotted with a poor grade is not considered for fresh order. In case of sufficient number of eligible vendors are not available in the existing vendor data base some new potential vendors are searched and invited to send their basic data highlighting the terms and conditions.

The information of all those vendors who respond positively with satisfactory terms and conditions is added to vendors' database. For effective vendor rating which leads to the identification of a good vendor an exhaustive set of data is collected form each vendor. The extent of vendor data details can be visualized by the contents of vendor database as shown in Table 5.3.

Table 5.3: Contents of Vendor Database

Vendor database
• Name and address – *Characters*
• Vendor code – *Alphanumeric*
• Contact years – *Numeric*
• Number of orders supplied – *Numeric*
• Late deliveries – *Numeric*
• Distance – *Numeric*
• Number of Rejections – *Numeric*
• List of materials – *Alphanumeric*
• Modes of transportation – *Characters*
• Terms of credit – *Characters(arrays)*
• Grade allotted to vendor – *Numeric*

Once the message is transferred form LIPA along with the list of items to be procured, it first identifies all the eligible vendors and then sends quotations to all eligible vendors

5.2.4 AGENT 4 – Quotation Receiving and listing Agent (QRA)

This agent accepts the responses of all the vendors to whom quotations have been sent and process the information submitted by them, in order to verify satisfaction of minimum terms and conditions and also submission of expected information which is necessary for performing vendor rating. If sufficient information is not provided than two possible actions are planned. One, to send a reminder to the vendor under consideration if, information about this vendor is available in the vendor database and has good ranking based on his performances, otherwise the second action is neglect the quotation of this vendor.

This agent compiles the available information and prepares the list of vendors who satisfy the above mentioned requirements for further processing and submit the list to VRA. It makes use standard formats which clearly specify and arrange systematically the necessary information for vendor rating.

5.2.5 AGENT 5 – Vendor Rating Agent (VRA)

This agent accepts the information form QRA by sending predefined message about the responses of all the eligible vendors to further process the information submitted by them. The primary objective of processing of the information is to select a suitable vendor for each item to be purchased. For this purpose, it performs vendor rating based on factors like cost, quality, delivery time and transportation cost. The final output is Selected Vendors List (SVL) (Fig. 5.2).

While performing the vendor rating, first some critical factors like proximity, reputation, flexibility, stability in market, expertise and subjective factors like plant administration, cost control machining capacity etc. are considered. Vendor fulfilling the minimum requirements of critical and subjective factors are further considered for inclusion in the suggested vendors list for evaluation. No comparison is made on the basis of these factors.

Final evaluation of vendors is carried out by using weight point plan with objective factors like quality, service (delivery), and price and transportation costs. Each factor will have a certain weight depending upon its importance such that total weight is 100%.

For each vendor the scores for each factor are calculated and then vendors' performance rate is determined.

Let S_p, S_q, S_d, and S_t represent the scores of a particular vendor for performance factors price, quality, service (delivery) and transportation costs

Similarly Let W_p, W_q, W_d, and W_t represent the weightings assigned to the performance factors price, quality, service (delivery) and transportation costs

Vendor Performance Rating for vendor under consideration is given by the sum of scores in all the performance factors i.e. $VPR = S_p + S_q + S_d + S_t$

The scores can be calculated as follows:

For price of the item

$$S_p = W_p \times LC/C_v$$

For quality if there are r% rejects then,

$$S_q = W_q (1 - r/100)$$

For service, if there are 'd' late deliveries

$$S_d = W_d (1 - 5 * d/100)$$

For Transpiration cost, lowest price offered is taken as standard.

$$S_t = W_t * \text{lowest price/actual price}$$

During the process of vendor selection for a particular material say M1 the total scores which are also referred as Vendor Performance Rating, VPR is calculated for each eligible vendor as shown in Table 5.4.

Table 5.4: Vendor Performance Rating Calculation

Vendor	Score for Price	Score for Quality	Score for Delivery	Score for Transportation	Total Score/VPR
V1	SP1	SQ1	SD1	ST1	T1
V2	SP2	SQ2	SD2	ST2	T2
V3	SP3	SQ3	SD3	ST3	T3
V4	SP4	SQ4	SD4	ST4	T4
…………	…………	…………	…………	…………	…………
VN	SPN	SQN	SDN	STN	TN

Vendor with highest Total score/Performance Rating is selected and it will be recommended to send the purchase order for this selected vendor. This agent presents its output as SVL (selected vendor list) and make available to other agents such as POGA.

5.2.6 AGENT 6 – Purchase Order Generation Agent (POGA)

This agent collects information from the VRA in the form of SVL about the selected vendors and generates purchase order for each of the item to be procured in a predefined form mentioning all the terms and conditions. Its output is presented as Purchase Orders (PO) (Fig. 5.1). For generating purchase ordered POGA employ a standard form developed for this purpose, The form precisely indicate the Material name, quantity, delivery date and terms and conditions of purchase.

The generated purchase orders are dispatched to the respective vendors with the help of their contact details. This agent is also responsible for obtaining the confirmation that the selected vendor has received the purchase order and has sent a message of acceptance of order.

5.2.7 AGENT 7 – Purchase Order Tracking Agent (POTA)

After dispatching the purchase orders to all the selected vendors POGA transfer this information to POTA for further action. Accordingly, using the information consisting of contact details of each selected vendors POTA perform a set of well defined actions to keep track of the purchase order and maintain continuous records about the status of purchase orders. This is necessary in order to ensure that the required quantities of material are received in stipulated time. If necessary this agents also perform the function of expediting of a purchase order.

5.2.8 AGENT 8 – Material Receipt & Stock Updating Agent (MRSUA)

This agent is also called as 'stores agent' and is responsible for collecting information about the receipt of ordered quantity, confirmation about its quality and updates the quantity of respective items in store. Once, a lot is received through a purchase order in the store then the lot is checked for both quantity and quality by a team of inspectors and submits a report to MRSUA about the suitability of the received lot. MRSUA immediately updates the available quantity in the store as per the quantity received and pass this information to the agents ICA and VUA for executing their functions. In particular it sends requisite recommendations to VUA for updating the vendor data base to assign suitable grade for the current vendor.

5.2.9 AGENT 9 – Vendor data Updating Agent (VUA)

Once a purchase order is executed successfully, MRSUA transfer the detailed information about the vendor involved in the current purchase order. If this vendor is new and supplying material first time to the organization, then the data about this vendor will be added to the vendor database with a suitable rank based on the extent to which the vendor under consideration is able to meet the terms and conditions specified in the purchase order. If the vendor involved is an existing vendor, then his rank will be updated based on his performance in executing the current purchase order.

5.2.10 AGENT 10 – Invoice Clearance Agent (ICA)

This agent act immediately after receiving message from MRSUA about the quantities of material added to store. It communicates with agents of financial aspect module and then recommends and guides to forward the invoice for payment as well confirm the payment of invoice. If the supplier has allowed certain time for the payments then this agent will obtain confirmation of such payments as and when it occurs. Once this agent executes its function the overall activities related to procurements of certain material will be completed.

The Logic for the purchasing process is depicted in flow chart shown in the Fig. 5.3.

Multi Agent Framework for Purchasing Process **139**

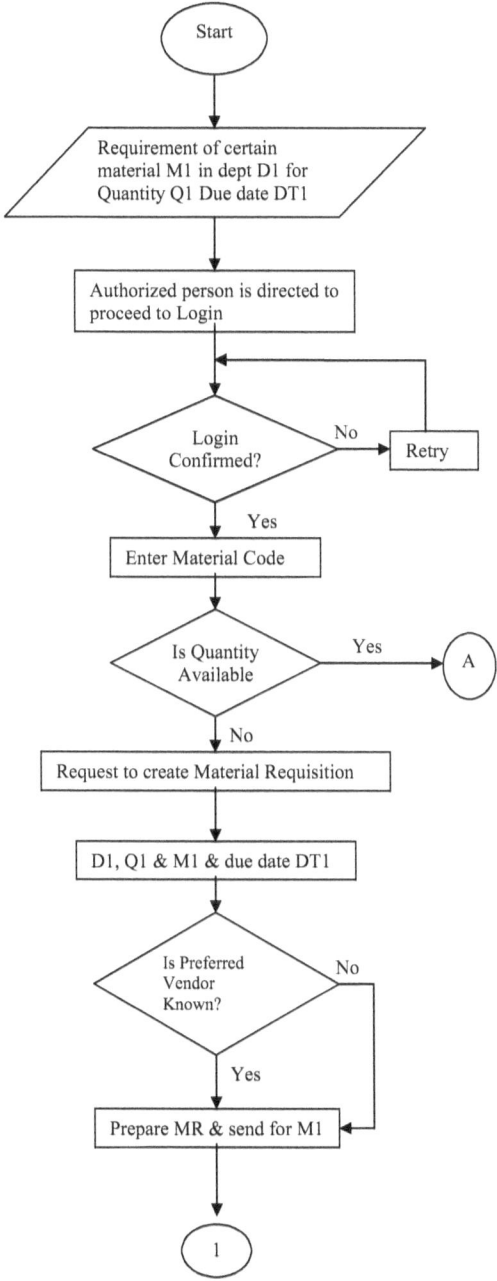

Fig. 5.3: Flow Chart Describing Logic of Executing Agents for Purchasing

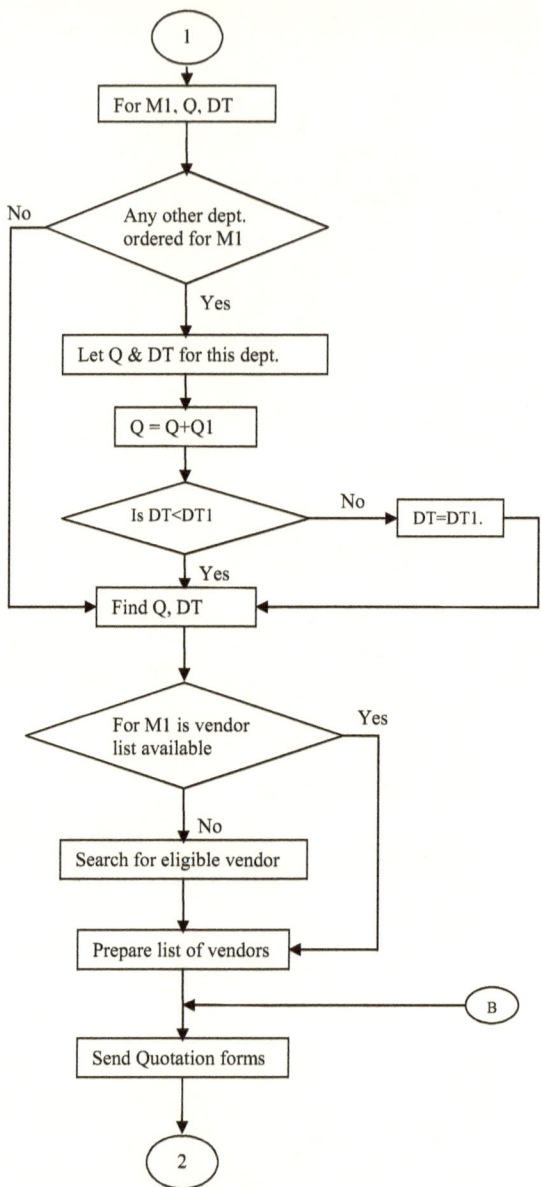

Fig. 5.3: (Continued)

Multi Agent Framework for Purchasing Process **141**

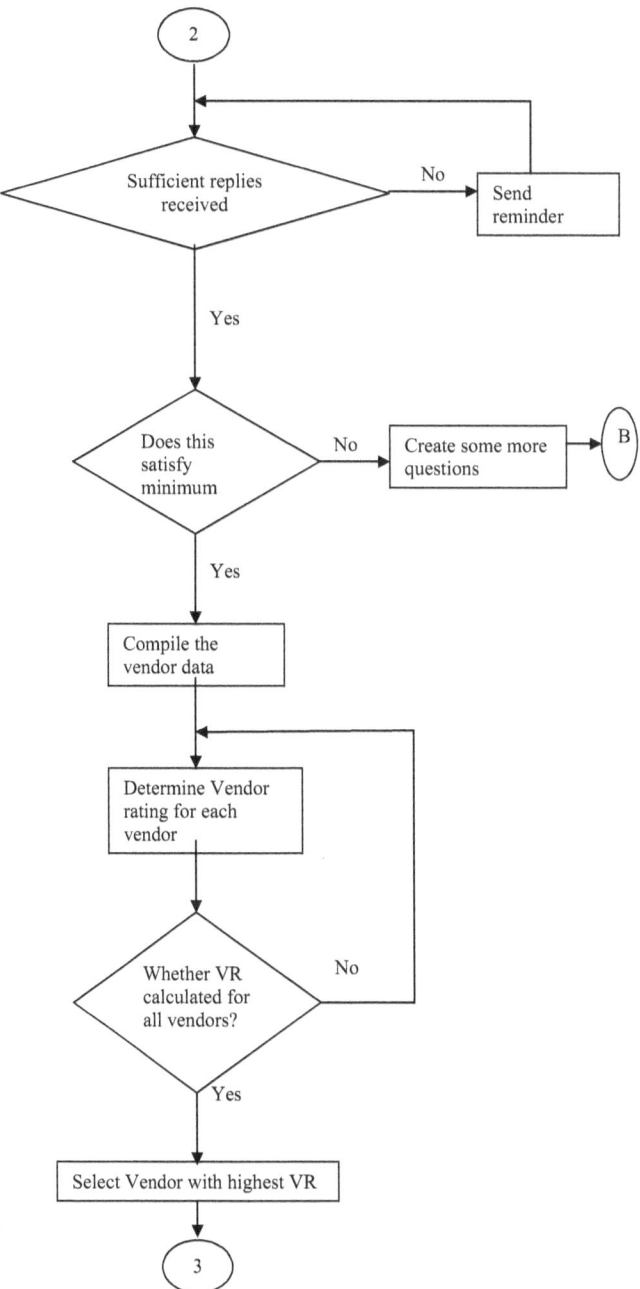

Fig. 5.3: (Continued)

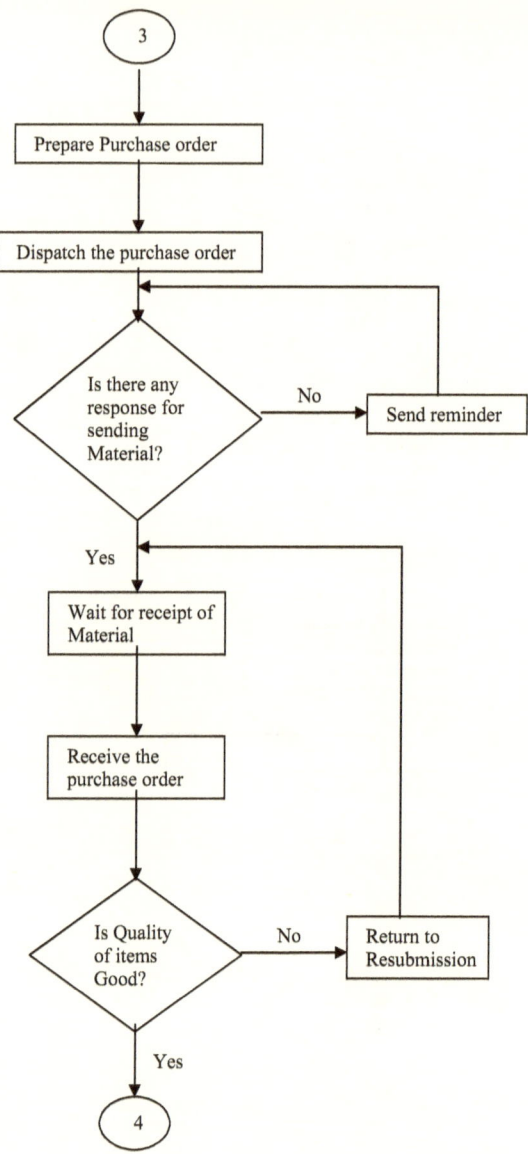

Fig. 5.3: (Continued)

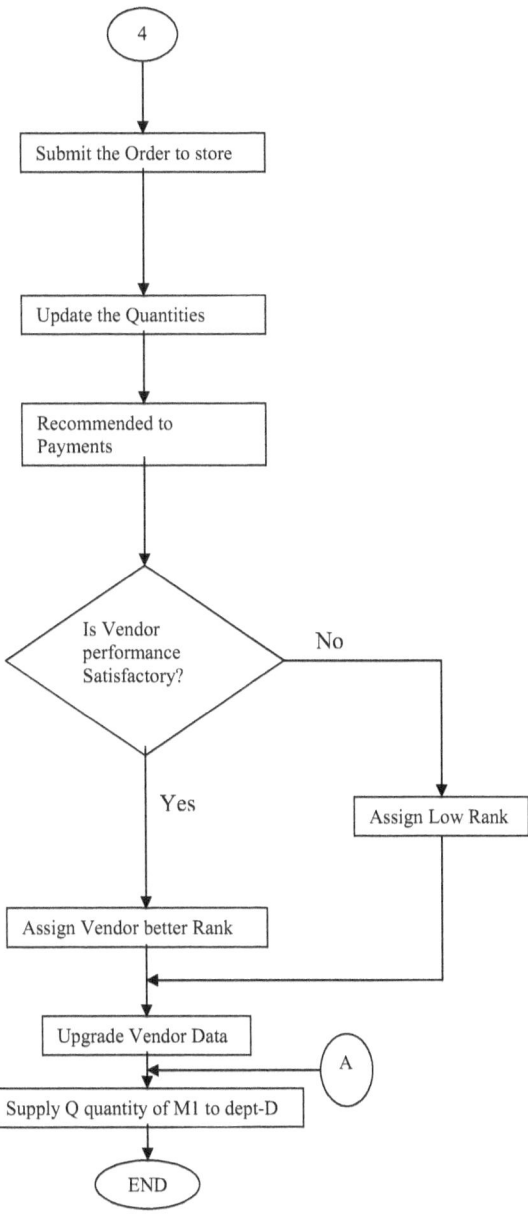

Fig. 5.3: (Continued)

The above flow chart depicts the logic for automation of the purchasing process in detail.

Chapter 6

Execution of Agents for Performing Activities Related to Purchasing

For a given master production schedule once, the MRP module performs the time phasing of requirements, then it lead to the generation of information about the inventory needs of individual departments/shops and sections. Each department or section identifies the list of items that are necessary to carry out its expected activities in given period of time and accordingly performs certain activities to ensure that requires quality and quality of material are available at right time at its disposal.

In every Organisation there will be an official who is responsible for ensuring that there is no shortage of any required inventory. He is also referred as the 'authorized user' who can access the storage database to check for availability of inventory, can order the inventory form store or can submit the material requisition as per the requirements of his department/section.

Let a department D1 has the requirement of raw inventory M1 in the nearest time bracket T1. Then the authorized user can check whether sufficient quantity of M1 is available in for the period T1 or not by logging in through login screen shown below in Fig. 6.1. For this purpose the authorized user has to execute **ACMRA (A**vailability **C**hecking **&** **M**aterial **R**equisition **A**gent) which present to him the log-in screen. The person has to enter the unique number assigned to him periodically.

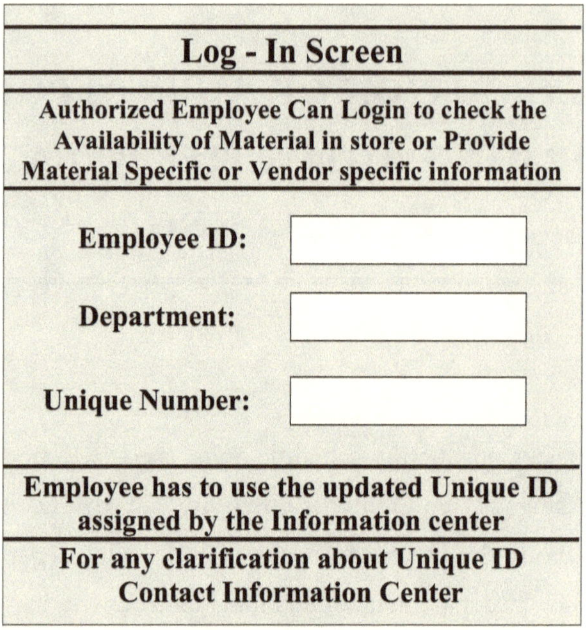

Log - In Screen
Authorized Employee Can Login to check the Availability of Material in store or Provide Material Specific or Vendor specific information
Employee ID: ☐
Department: ☐
Unique Number: ☐
Employee has to use the updated Unique ID assigned by the Information center
For any clarification about Unique ID Contact Information Center

Fig. 6.1: Log-In Screen

Once the employee successfully Logs-in then the welcome screen present him the following three options for interacting with the storage database (refer Fig. 6.2).

- To view the available quantity of a specific material by specifying its material code or the list of available quantities in the storage database,
- To provide certain data about an existing vendor or a new vendor, which may be significant for the future purchasing activities and also to update vendor database

- To submit certain material specific information based on some feed back obtained form shop floor personnel which may be significant for future purchasing activities

The user is directed to clicks on the specified points "A" or "V" or "M" for choosing any of the above mentioned options respectively.

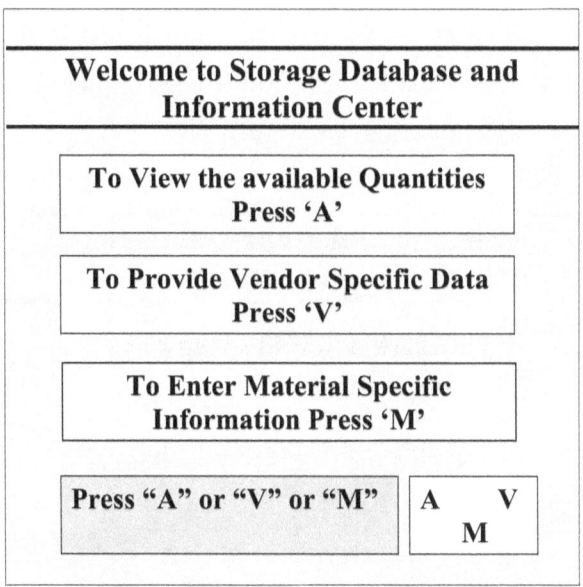

Fig. 6.2: Welcome Form for Storage Database

Once, the authorized user selects the first option by clicking on "A" then a new form is presented to the user with the following two choices as shown in the Fig. 6.3.

- To view exhaustive list available raw inventory,
- To view the available quantity for a specific material…

The form shown in Fig. 6.3 allow the user to choose any one of the above mentioned options by clicking on the suitable key i.e. "M" for a specific material and "A" for the list of all the available items with quantities.

If user clicks for choice for checking the availability of a specific raw material, then the corresponding form as shown in Fig. 6.4 will be displayed and the user is prompted to mention his request by entering the Material name and Material code. If needed the form provides a link to get the details about the coding scheme of all the material available in the store. The form also has a provision to invite the user to verify the available quantity of another specific raw material. The user can avail this facility in order to obtain the information about the available quantity for a group of specific items one after another.

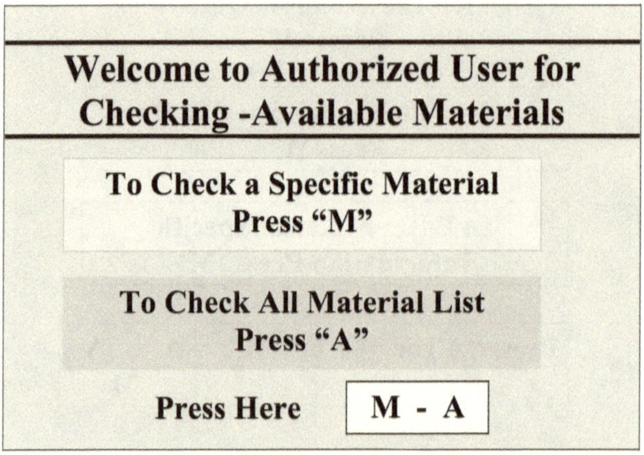

Welcome to Authorized User for Checking -Available Materials

To Check a Specific Material Press "M"

To Check All Material List Press "A"

Press Here M - A

Fig. 6.3: Welcome Form for Checking Available Material

For Verification of a Specific Material

Enter Following

Material Name	
Material Code		Click Here

If You Are You Interested To Check for other Material Press "Y"   Y   N

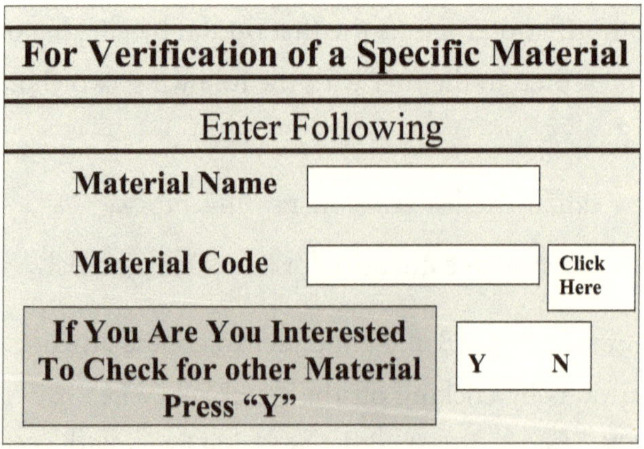

Fig. 6.4: Form for Checking Availability of a Specific Material

Once the user enters the material code and material name and confirms by clicking at the suitable box to receive the required information, then ACMRA verifies the material code and material name and if they are part of storage database, it displays the details available about the material under consideration through the form shown below as Fig. 6.5

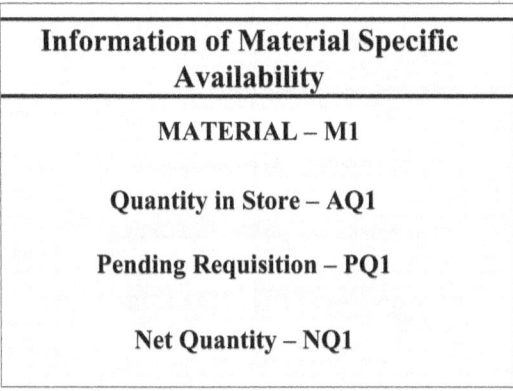

Fig. 6.5: Form Displaying the Details of Specific Material Form Storage Database

It should be noted that if the user makes improper entries for material name and code then, ACMRA guides the user to use the link for obtaining the information about list of materials and their codes and reselect the form shown in Fig. 6.4.

In the above form the quantity in store indicates the actual quantity available in store at this moment. There can be some pending orders which are submitted by other authorized users through the current procedure. It means the authorized users from other departments/sections which are in need of this material M1 have already verified and confirmed the availability of this material and requested for supply of the same. However, these requested quantities are yet to be supplied to the respective departments/sections. On priority basis as these quantities are requested earlier, first the delivery of material is carried out for these orders. Hence, for the current authorized user the available quantity will be the quantity that may remain in storage after satisfying the pending requisitions. This quantity will be shown as the 'net quantity'.

In order to get a complete list of all the available material in the store, if the user chooses the option of pressing "M" on the form shown in Fig. 6.4, then the form shown in Fig. 6.6 will by displayed to presents the required information.

This is the list of all the Items/ Materials with Available Quantity

Material Name	Material code	Available Quantity	Pending Requisition	Net Quantity
Material –1	M1	AQ1	PR1	NQ1
Material – 2	M2	AQ2	PR2	NQ2
Material – 3	M3	AQ3	PR3	NQ3
Material – 4	M4	AQ4	PR4	NQ4
Material – N	MN	AQN	PRN	NQN

To Check a Specific Material Click on its code

To Obtain a list for a set of Selected Materials Click on respective codes

Fig. 6.6: Form for Checking All Available Material List

The above form not only shows the details of all the material available in stock but also provides with options for getting details of specific material by clicking on the code of that material on the form or the details of a selected group of materials by continuously pressing on the necessary codes.

The two forms shown in Fig. 6.5 and Fig. 6.6 provide an insight into the storage database and sufficient information about the Material Available in the stock for and further give the clue to the authorized user in order to take appropriate action to ensure the supply of required quantity of material for smooth production activities in the respective department or section. If the net quantity displayed is sufficient for fulfilling the expected demand of the user then ACMRA permits the user to choose the order form shown below in Fig. 6.7

```
┌─────────────────────────────────────┐
│      Order Form - From Store        │
│         Enter the following         │
│                                     │
│   Material Name  [          ]       │
│                                     │
│       Code       [          ]       │
│                                     │
│     Quantity     [          ]       │
│                                     │
│   Due date:      [ DD/MM/YYY ]      │
│                                     │
│    Department    [          ]       │
│                                     │
│    [ Click Here for Submitting ]    │
│                                     │
└─────────────────────────────────────┘
```

Fig. 6.7: Form for Submitting Request for Supply of Material from Store

Choosing this form now the user can enter the name of code of the material required for his/her department/section and also specifies the quantity and due date of requirements along with the name of the department. Once the user submits this form with all details ACMRA converts this request into pending requisition and add it to the storage data base. It also assures the user about the supply of required material at right time as per the sue date requirement.

If the user finds that net quantity in the store is insufficient to meet his requirements, then he/she can request the ACMRA to present the materiel requisition form (refer Fig. 6.8) in order to submit his demand for the material and requesting for procurement of the required material by the purchasing department.

Name of the Organization (printed)					
Material Requisition Form					
Department:			MREP No:		
			Date:		
S.No	Material Name	Material Code	Item Specification	Quantity	Preferred Vendor
Items are required on or Before:		DD /MM /YY			
Authorized Person (Name & ID)			Name of head of the Section		
Remarks If any					
Budget requirements (For purchase section)					
Budget Estimated			Authorized Officer		
Details of Budget Approval and Purchase Section Approval					

Fig. 6.8: Material Requisition Form

ACMRA displays the material requisition form and prompt the user to enter the following data for processing his request:

- Name of the department
- Name and code of the material
- Specifications of the material
- Required quantity
- Date on which items are required

The user is also permitted to enter the details of preferred vendor and any other significant information as 'remarks' which may support the purchasing of material under consideration.

Once, the user makes all entries in the material requisition form and submits it for consideration, and then ACMRA makes estimation of budgeting requirements for the requested requisition and obtains confirmation and approval for official in charge for the purpose. Finally, it enters all the details of budget requirements in appropriate box and with a formal approval submit the material requisition form to LIPA for further processing.

As highlighted above the welcome form of the storage database (refer Fig. 6.2) permits the authorized user to have an access to storage data for checking the availability of raw inventory. Apart from this facility the user also enjoys the provisions through which he can send information about a vendor or a material to storage database by selecting the choices "V" or "M" respectively.

If the user selects the option "V", then ACMRA presents the form shown in Fig. 6.9 and permit the user to enter data about a new vendor or any positive or negative remarks about an existing vendor based on shop floor activities.

Similarly the user can select option "M" and obtain the form shown on Fig. 6.10 which he can use to send information about either quality or quantity problem of a specific material.

Welcome to Authorized user
Enter Vendor Specific Data

Name:

Address:

Materials:

Remarks:

Fig. 6.9: Form for Entering Vendor Specific Data by Authorized User

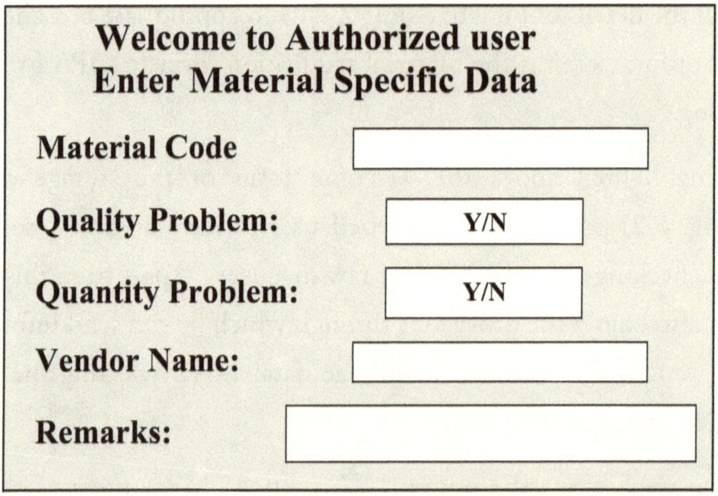

Welcome to Authorized user
Enter Material Specific Data

Material Code

Quality Problem: Y/N

Quantity Problem: Y/N

Vendor Name:

Remarks:

Fig. 6.10: Form for Entering Material Specific Data by the Authorized User

ACMRA groups all the approved material requisition forms submitted in a given time and submit them to LIPA for further action. LIPA compiles the information submitted ACMRA and prepares the consolidated list of items for purchasing in the near future. If a material is requested by more than one department than the due date is determined by the earliest required date among the dates specified by different departments/sections. LIPA arranges this list in a form as shown in Fig. 6.11 and submit it to QSA to proceed for the purchasing of materials.

This is the list of Items to be Purchased

Material Name	Material code	Quantity to Purchase	Due Date For purchase
Material –1	M1	Q1	DD/MM/YY
Material – 2	M2	Q2	DD/MM/YY
Material – 3	M3	Q3	DD/MM/YY
Material – 4	M4	Q4	DD/MM/YY
Material – N	MN	QN	DD/MM/YY

Fig. 6.11: Form Displaying List of Material to Be Purchased

The major function of QSA is to send the invitation to the eligible vendors to send their quotation for the set of materials required for various departments. Before performing this function it also performs the function of identifying the set eligible vendors for all the materials. For this it browses the vendor database to identify all those vendors who have been assigned higher grades based on their effectiveness in supplying previous orders. The vendor database provides information about each existing vendor in a form as shown below in Fig. 6.12.

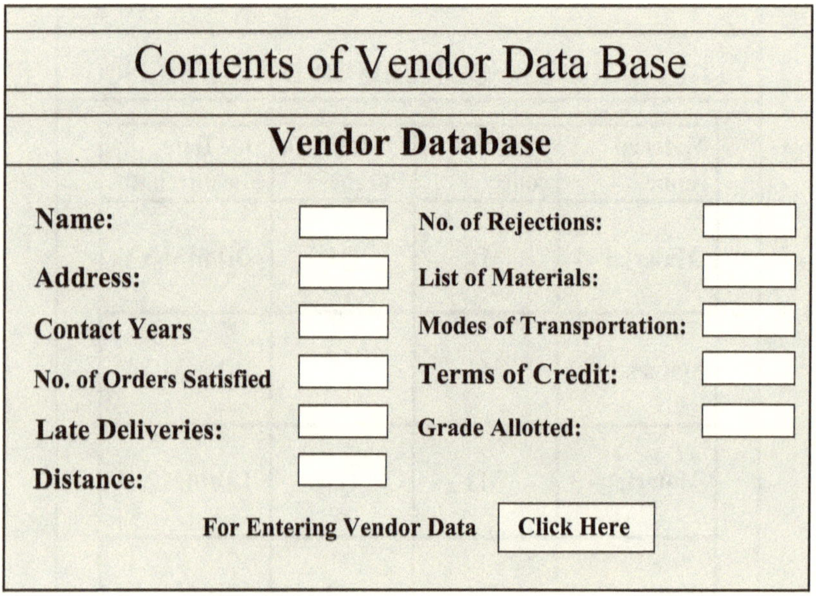

Contents of Vendor Data Base

Vendor Database

Name:		No. of Rejections:	
Address:		List of Materials:	
Contact Years		Modes of Transportation:	
No. of Orders Satisfied		Terms of Credit:	
Late Deliveries:		Grade Allotted:	
Distance:			

For Entering Vendor Data — Click Here

Fig. 6.12: User Screen for Vendor Database

If sufficient number of eligible vendors is not identified in the existing database then QSA follow certain well defined activities to obtain information about new vendors who satisfy the minimum requirements as per the norms of the organization. After ensuring that sufficient vendors list is prepared then QSA prepares the list of materials along with vendors identified in the form as shown in 6.13.

This is the list of Items with vendors		
Material Name	Material code	Vendor details
Material –1	M1	V1,V3,V5,V8
Material – 2	M2	V2,V4,V5,V6,V7
Material – 3	M3	V1,V3,V7MV8
Material – 4	M4	V2,V4,V5,V8
Material – N	MN	V1,-------V8

Fig. 6.13: Screen for Material with Vendors

To each vendor who has been identified in the above form, QSA sends a request letter inviting the vendor to send the quotation for all those materials which are specified against this vendor. For sending this invitation QSA utilize the form shown in Fig. 6.14. Each vendor is requested to use the form shown in Fig. 6.1.5 for submitting the quotation. The vendors are specifically informed to specify the following information precisely for each material.

- Price of the item
- Lead time for supply of items
- Quality in terms expected % of defectives in a lot
- Mode of transportation and
- Cost of transportation

Invitation for Quotation

To,
Vendor Name :

Address:

Subject : Request for sending Quotation for the material listed Below

Reference No :

Sl.No.	Item Name	Item Code	Quantity	Due Date
1.	IN1	IC1	IQ1	DD/MM/YYYY
2.	IN2	IC2	IQ2	DD/MM/YYYY
3.	IN3	IC3	IQ3	DD/MM/YYYY
N.	INN	ICN	IQN	DD/MM/YYYY

You are requested to send your Quotation for the above Items at the earliest specifying following

Price   **Lead Time**

Quality

Transportation Modes and Cost

Note : The terms and conditions of the Purchasing are attached as file

Signature

(Authorized Purchasing officer)

Fig. 6.14: Requisition Form for a Quotation

Quotation for Material

Name of the Organization :

Address:

Reference No   **Date**

Reference No of Request letter:							
Phone: Fax:							
S. No	Material Name	Code	Description	Quantity	Price	Total	Lead Time
I HAVE READ THE ABOVE ORDER. I AGREE AND UNDERSTAND ALL THE ABOVE SPECIFICATIONS. I UNDERSTAND IT IS MY RESPONSIBILITY TO MEET ALL CODES; LEAD TIMES ARE APPROXIMATE AND DEPEND UPON SUPPLIER(S). Dated:_____ Signed_____					Total Cost		
					Tax:		
					TOTAL Cost of Quotation :		

Distance

Mode of Transportation

Fig. 6.15: Quotation Form

After sending the invitation letters to various vendors identified as eligible vendors, QSA transfer this information about the vendors to QRA, which in turn accepts the quotations submitted by the eligible vendors and checks for minimum terms and conditions which are necessary to perform Vendor rating. All those vendors who are found to be satisfying minimum required conditions for vendor rating are called as qualified vendors. For a particular material if sufficient numbers of qualified vendors are not available then,

QRA pass a message to QSA requesting it to make arrangements for the quotations of few more vendors.

Once, sufficient numbers of qualified vendors list is prepared QRA arranges this information in a standard format in which for each material a set of qualified vendors are specified. QRA sends this list to VRA for performing vendor rating.

After receiving material wise vendors list VRA perform the vendor rating based on the following factors l

- Quality of the material,
- Cost of the materials,
- Transportation cost
- Delivery aspects.

For delivery aspects either the delivery time specified by the vendors in their quotation or the information about number of delayed deliveries forms the vendor database (if all the qualified vendors are existing vendors). Each of these factors may be treated with equal or different degrees of importance. The importance of each factor is specified through certain weightage assigned to that factor such that total weightage will be 100.

VRA determines and assigns the weightages to different factors considered for vendor rating, depending upon the purchasing policies of the organization. Further, it considers materials one after other and for each, material calculates Total score/Performance Rating for each qualified vendor as displayed in form shown in Fig. 6.16.

After performing vendor rating for each material and making the final selection of a vendor for that material VRA prepares a list of selected vendors and presents it through the list SLV as shown in Fig. 6.17. Hence, SLV becomes the prime output of VRA, which will be submitted to POGA to carry out the further steps for purchasing of materials under consideration. SLV will be submitted to POGA by VRA with recommendations to place the purchase order for this selected vendor.

With SLV as its major input, POGA generates purchase order for every selected vendor in the format shown in Fig. 6.18, highlighting clearly characteristics of materials, their quantity and due date requirements, mode of transportation, price aspects and also other terms and conditions of purchasing.

Compiled Vendor Rating for Material M1						
Vendor	Address	Score for Quality	Score for Cost	Score for Distance	Score for Transportation	Total score
V1	A1	SQ1	SC1	SD1	ST1	TS1
V2	A2	SQ2	SC2	SD2	ST2	TS2
V3	A3	SQ3	SC3	SD3	ST3	TS3*
-------	-------	------	-----	------	-----	------
VN	A-Z	SQN	SCN	SDN	STN	TSN
Vendor V3 is selected since this vendor has performed better in terms of highest Total score*						

Fig. 6.16: Form Displaying Vendor Rating Data for a Specific Material

Selected Vendors list (SLV)		
Material Name	Material code	Vendor details*
Material –1	M1	V2
Material – 2	M2	V4
Material – 3	M3	V1
Material – 4	M4	V4
Material – N	MN	VN

*The details will be obtained from the vendor database

Fig. 6.17: Screen for Material with Vendors

Purchase Order for Material

Name Company:

Vendor Address:

Shipping Address:

Item Code	Description	Quantity	Unit Price	Total
			Subtotal	
			Tax Rate	
			Tax	
			TOTAL	

Mode of Transportation:

Terms and Conditions:

Fig. 6.18: Purchase Order

POGA is not only responsible for preparing the purchase orders in the format shown above but also to dispatch the purchasing orders to respective vendors. After that it sends the detailed information about all the purchase orders to POTA which is capable of performing pre defined activities to maintain the status of all the purchase orders.

POTA contacts the supplier for either for updating the delivery schedule or to reassess that schedule of delivery of items is maintained and ensures the securing of the quality and timely delivery of goods and components. For this it employs the enquiry form displayed in Fig. 6.19 to communicate with the vendors.

```
┌─────────────────────────────────────────────────┐
│ Organization / Company Name: [        ]         │
├─────────────────────────────────────────────────┤
│         Reminder for status of                  │
│            Purchase order                       │
├─────────────────────────────────────────────────┤
│ To Vendor: [              ]                     │
├─────────────────────────────────────────────────┤
│ The following purchase order has been sent for  │
│ the materials listed below                      │
│ Purchase order Details : [            ]         │
└─────────────────────────────────────────────────┘
```

Sl. No.	Item Name	Item Code	Quantity	Due Dates Remarks
1.	IN1	IC1	Q1	DD/MM/YYYY
2.	IN2	IC2	Q2	DD/MM/YYYY
3.	IN3	IC3	Q3	DD/MM/YYYY
----	---------	------	--------	--------------
N.	INN	ICN	QN	DD/MM/YYYY

<u>You are requested to respond immediately to indicate the Status of executing the above Purchase order</u>.

Signature

(Purchasing In charge.)

Fig. 6.19: Form for Expatiation

When a purchased order is executed successfully the quantity supplied by the respective vendor will be received in the storage and the concerned people make use of MRSUA. The MRSUA allow them to use the form shown in Fig. 6.20 to prepare the report about the acceptance/rejection of the lot received. The official personnel for inspection makes a complete inspection of the lot by applying the quality control methods to decide about the acceptability of the lot. The result of the inspection of lot is expressed though the following form along with suitable recommendations for further action.

Form for Clearance Report / Rejection Note

Report / Note No. [   ]   **Date:** DD/MM/YYYY

Sl. No.	Order No.	Material Name	Item Code	Supplier	Remarks

The above mentioned Materials have been Accepted [   ] **/ Rejected** [   ] **through Inspection.**

Details of Examination:	Reasons for Rejection (if rejected) :
Goods received by Signature (Stores Keeper)	**Inspected by:** Signature (Inspection Department)

Fig. 6.20: Form for Acceptance or Rejection of Materials Received

If the lot is recommended for acceptance then it is officially received by the stores personnel and submits the relevant information by making entries in the form shown in Fig. 6.21. Once this form is submitted by stores personnel, MRSUA automatically updates the storage database by enhancing the available quantities suitably. In addition MRSUA sends message to the VUA to update the vendor database by reassigning the new ranks to the vendor under consideration.

Form for the Material Received

Delivery point:		Purchase Order No.	
From: (Vendor)		Vendor's Address:	

Quantity received	Description of Material	Condition of Goods	Grade

Counted and Inspected by		Approved by	

Remarks For Vendor	

Fig. 6.21: Form for Material Received

MRSUA also sends message to ICA to perform a set of well defined activities in order to communicate the agents of financial aspect module with recommendations for timely payments to the vendors who have executed the current purchase orders satisfactorily.

Chapter 7

Results and Discussions

The discussions presented in the earlier chapters indicates that in the current work an attempt has been made to apply the concepts of agent technology, in particular multi agent systems to develop a frame work for the flow of information between various unit of supply chain formed for a manufacturing organization. The outcome of this work leads to the effective sharing of information not only with in the organization (intra organization activities) but also between the organizations (inter organizational activities) involved din the supply chain. After presenting a modular frame work, for the over all flow of information for supply chain activities, specific agents have been developed for the following functional areas.

- Procurement of materials and components
- Material requirements planning
- Process planning
- Production Scheduling

For validation and demonstration of the work efforts were carried out to collect and employ some real life problems related to those manufacturing organizations which are making attempts or planning to make attempts to apply supply chain strategies for their activities. However due to the policies of industrial organizations it was not possible to obtain data describing the real life problem which can be used for the proposed work. As an alternative, after collecting the relevant data form a manufacturing organization, the data has been restructured to avoid the revealing of the identity of the organization. It may be noted that in this process some amount of accuracy and reliability of data may be lost.

With the restructured industrial data a case study has been formulated and product structure of an example part is defined through its bill of material as shown in the Fig. 7.1. With the given master production schedule for the part (refer Table 7.1) time phased requirements for its components are determined through manual calculations and presented thorough various tables (Table 7.2). Form these requirements period wise list of items to be manufactured and to be purchased to satisfy the given master schedule are identified and presented in Tables 7.3 and 7.4 respectively.

Table 7.1: MSP for Example Part

Time-Week	10	11	12
Quantity	100	150	100

Assumptions made are: on hand inventory for all Items = 0.

Table 7.2: Time Phased Requirement of Different Components of Example Part

Table 7.2.1: Planned Requirement for Chair (C)

Time-Week	8	9	10	11	12
Requirement			100	150	100
Inventory Stock		0	0	0	0
Net Requirement			100	150	100
Planned Order release		100	150	100	

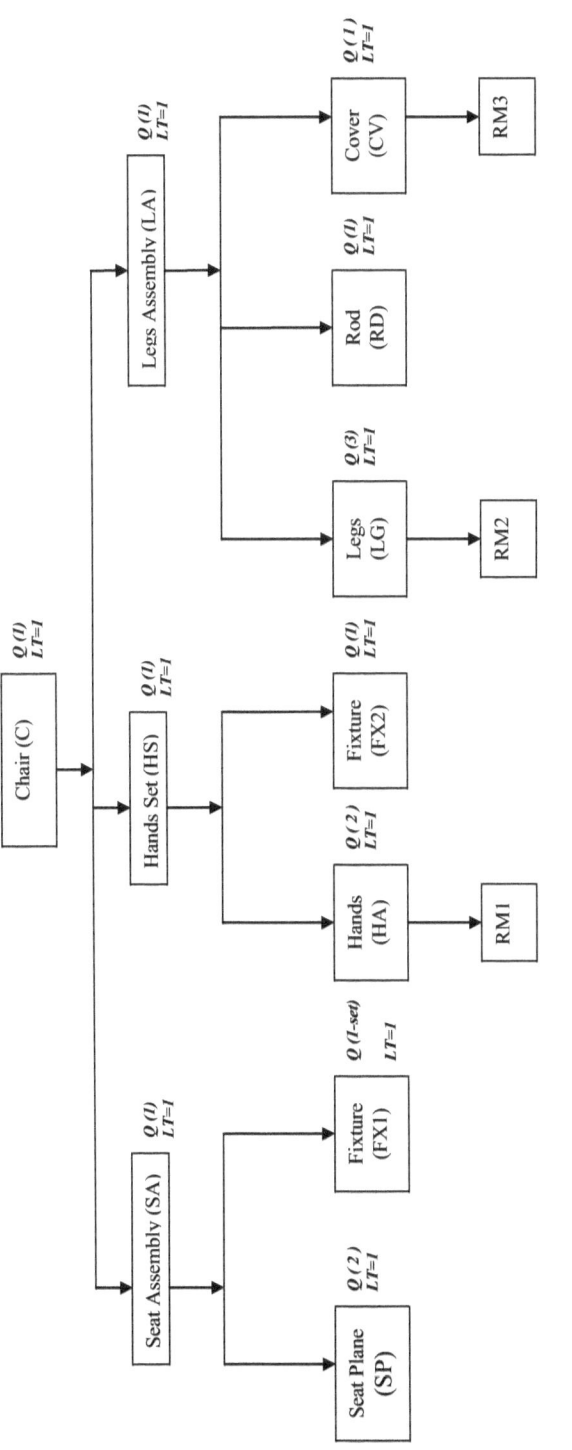

Fig. 7.1: BOM for Example Part

Q – Quantity, LT – Lead Time, RM – Raw Material

Table 7.2.2: Planned Requirement for Seat Assembly (SA)

Time-Week	8	9	10	11	12
Requirement		100	150	100	
Inventory Stock		0	0	0	
Net Requirement		100	150	100	
Planned Order release	100	150	100		

Table 7.2.3: Planned Requirement for Hands Set (HS)

Time-Week	8	9	10	11	12
Requirement		100	150	100	
Inventory Stock		0	0	0	
Net Requirement		100	150	100	
Planned Order release	100	150	100		

Table 7.2.4: Planned Requirement for Legs Assembly (LA)

Time-Week	8	9	10	11	12
Requirement		100	150	100	
Inventory Stock		0	0	0	
Net Requirement		100	150	100	
Planned Order release	100	150	100		

Table 7.2.5: Planned Requirement for Seats Plane (SP)

Time-Week	7	8	9	10	11
Requirement		200	300	200	
Inventory Stock		0	0	0	
Net Requirement		200	300	200	
Planned Order release	200	300	200		

Results and Discussions 171

Table 7.2.6: Planned Requirement for Fixture (FX1)

Time-Week	7	8	9	10	11
Requirement		100	150	100	
Inventory Stock		0	0	0	
Net Requirement		100	150	100	
Planned Order release	100	150	100		

Table 7.2.7: Planned Requirement for Hands (HA)

Time-Week	7	8	9	10	11
Requirement		200	300	200	
Inventory Stock		0	0	0	
Net Requirement		200	300	200	
Planned Order release	200	300	200		

Table 7.2.8: Planned Requirement for Fixture Set 2 (FX2)

Time-Week	7	8	9	10	11
Requirement		200	300	200	
Inventory Stock		0	0	0	
Net Requirement		200	300	200	
Planned Order release	200	300	200		

Table 7.2.9: Planned Requirement for Legs (LG)

Time-Week	7	8	9	10	11
Requirement		300	450	300	
Inventory Stock		0	0	0	
Net Requirement		300	450	300	
Planned Order release	300	450	300		

Table 7.2.10: Planned Requirement for Rod (RD)

Time-Week	7	8	9	10	11
Requirement		100	150	100	
Inventory Stock		0	0	0	
Net Requirement		100	150	100	
Planned Order release	100	150	100		

Table 7.2.11: Planned Requirement for Cover (CV)

Time-Week	7	8	9	10	11
Requirement		100	150	100	
Inventory Stock		0	0	0	
Net Requirement		100	150	100	
Planned Order release	100	150	100		

Table 7.2.12: Planned Requirement for RM1

Time-Week	6	7	8	9	10
Requirement		200	300	200	
Inventory Stock		0	0	0	
Net Requirement		200	300	200	
Planned Order release	200	300	200		

Table 7.2.13: Planned Requirement for RM2

Time-Week	6	7	8	9	10
Requirement		300	450	300	
Inventory Stock		0	0	0	
Net Requirement		300	450	300	
Planned Order release	300	450	300		

Table 7.2.14: Planned Requirement for RM3

Time-Week	6	7	8	9	10
Requirement		100	150	100	
Inventory Stock		0	0	0	
Net Requirement		100	150	100	
Planned Order release	100	150	100		

This case study is then submitted to the agents grouped under MRP module and the out put generated by these agents is compared with the above calculations for verification. It is found that the result is satisfactory. Then as per the time phased requirements (refer Table 7.2), individual departments are made to submit their material requisitions through agent, ACMRA. From ACMRA this information was transferred to LIPA which prepared the list of items to be purchased and presented the list through the prescribed form as shown in Fig. 7.2 for further needful action by QSA. This list that is generated by the LIPA is compared with the list which is prepared manually (refer Table 7.4) and it is found that the calculation perfectly match with each other.

Table 7.3: List Items to Be Manufactured

Time-Week	Status of Item	Item	Quantity	Vendor
07	Manufactured	CV	100	
07	Manufactured	LG	300	
07	Manufactured	HA	200	
08	Manufactured	CV	150	
08	Manufactured	LG	450	
08	Manufactured	HA	300	
08	Manufactured	LA	100	

Contd...

Time-Week	Status of Item	Item	Quantity	Vendor
08	Manufactured	HS	100	
08	Manufactured	SA	100	
09	Manufactured	CV	100	
09	Manufactured	LG	300	
09	Manufactured	HA	200	
09	Manufactured	LA	150	
09	Manufactured	HS	150	
09	Manufactured	SA	150	
09	Manufactured	C	100	
10	Manufactured	LA	100	
10	Manufactured	HS	100	
10	Manufactured	SA	100	
10	Manufactured	C	150	
11	Manufactured	C	100	

The list of items to be manufactured which is listed below in Table 7.4 will become the major input for the agents of scheduling module.

Table 7.4: List of Items to Be Purchased

Time-Week	Status of Item	Item	Quantity	Vendor
06	Purchased	RM3	100	
06	Purchased	RM2	300	
06	Purchased	RM1	200	
07	Purchased	RM3	150	
07	Purchased	RM2	450	
07	Purchased	RM1	300	

Time-Week	Status of Item	Item	Quantity	Vendor
07	Purchased	RD	100	
07	Purchased	FX2	200	
07	Purchased	FX1	100	
07	Purchased	SP	200	
08	Purchased	RM3	100	
08	Purchased	RM2	300	
08	Purchased	RM1	200	
08	Purchased	RD	150	
08	Purchased	FX2	300	
08	Purchased	FX1	150	
08	Purchased	SP	300	
09	Purchased	RD	100	
09	Purchased	FX2	200	
09	Purchased	FX1	100	
09	Purchased	SP	200	

Considering the list of items to be manufactured which is compiled below (refer Fig. 7.2), QSA acts further to identify and prepare the list of all eligible vendors and send quotations to these vendors. For this QSA scans the existing vendor database and also collecting data about new vendors through the form as shown in Fig. 7.3. QSA verified the data provided by individual vendors and separated all the vendors who satisfy the minimum terms expected by the organization. QSA designated these vendors as eligible venders and sent request letters for sending quotations to all the eligible vendors.

This is the list of Items to be Purchased

Material Name	Material code	Quantity to Purchase	Due Date For purchase
Cover	RM3	100	6th week
Legs	RM2	300	6th week
Hands	RM1	200	6th week
Cover	RM3	150	7th week
Legs	RM2	450	7th week
Hands	RM1	300	7th week
Rod	RD	100	7th week
Fixture Set 2	FX2	200	7th week
Fixture 1	FX1	100	7th week
Seat Plane	SP	200	7th week
Cover	RM3	100	8th week
Legs	RM2	300	8th week
Hands	RM1	200	8th week
Rod	RD	150	8th week
Fixture Set 2	FX2	300	8th week
Fixture 1	FX1	150	8th week
Seat Plane	SP	300	8th week
Rod	RD	100	9th week
Fixture Set 2	FX2	200	9th week
Fixture 1	FX1	100	9th week
Seat Plane	SP	200	9th week

Fig. 7.2: Form Displaying List of Material to Be Purchased

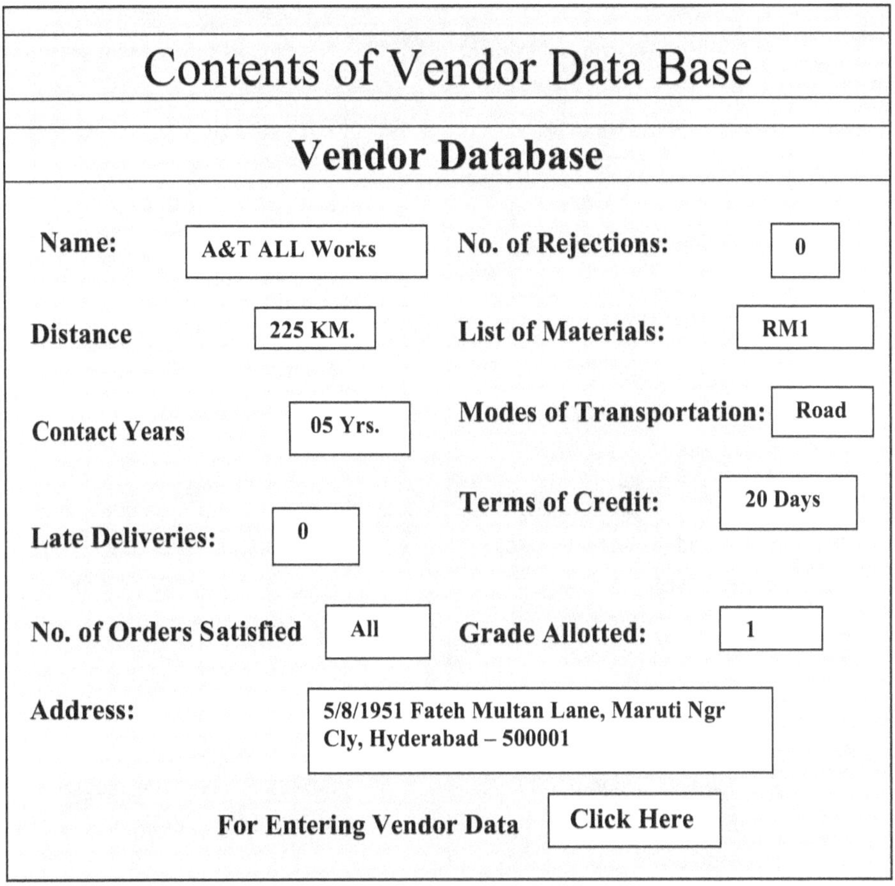

Contents of Vendor Data Base

Vendor Database

Name:	A&T ALL Works	**No. of Rejections:**	0
Distance	225 KM.	**List of Materials:**	RM1
Contact Years	05 Yrs.	**Modes of Transportation:**	Road
Late Deliveries:	0	**Terms of Credit:**	20 Days
No. of Orders Satisfied	All	**Grade Allotted:**	1
Address:	5/8/1951 Fateh Multan Lane, Maruti Ngr Cly, Hyderabad – 500001		

For Entering Vendor Data — Click Here

Fig. 7.3: Screen for Vendor Database

The list of vendors who are included along with their details in the vendor database is presented in the Table 7.5.

Table 7.5: List and Address of Vendors

Vendor Code	Name and Address of Vendor	Vendor Code	Name and Address of Vendor
V1.	A&T ALL Works 5/8/1951 Fateh Multan Lane Maruti Ngr Cly Hyderabad – 500001.	V2	Excellent I Systems F-13/B/1 Phase-8, Road No 56 Industrial Area, Hyderabad. – 500055.

Contd...

Vendor Code	Name and Address of Vendor	Vendor Code	Name and Address of Vendor
V3.	Square Table Products, Choukara palli, 49, 11th S Main, 12th Cross, Bangalore – 560013.	V4.	Rameshwara Eng. Works, 19-9-326, near Rudrayya High School, Ramnagar, Hyderabad, Andhra Pradesh India.
V5.	Sushmit Industries, K.S.M. I. Industrial Estate 25/9, 5th Cross, Janamanam, Bengaluru, Karnataka India.	V6.	Sai Sreepaad Industries, No. 7-9-456, Edla Bazar, Near City Play Ground, Hyderabad, Andhra Pradesh.
V7.	Lion Seating System, 23-9-562/42, maulali, Badrapur, Hyderabad, Andhra Pradesh India.	V8	Shivendra Enterprises 45 Sri nagar Colony Main Road, Hyderabad.
V9.	Maheshwari Industries, Plot No E-29 Phase-10 I D A, Ameerpur Ameerpur, Hyderabad.	V10.	Indian Furnitures, 8-3-234/1 Hanuman Nagar, Main Road, Hyderabad.
V11.	National Industries, Plot No 45 Door No 23/9, Near Nampalli School, Kagapunjar Nagar, Nungabakkam, Chennai – 600012.	V12.	Elegant Furnitures, No 7, 9a/50, Near Chella Koil, Dharur, Chennai – 600123.
V13.	Seating Solutions Co. No:104, Near traf Signal, Bluehills Road, New Kamakshi Puram, Chennai – 600003.	V14.	Dandapani Seaters, No 34, Opp To Sackura theatre, Jairam street, Kovur, Chennai – 600086.

Contd...

Vendor Code	Name and Address of Vendor	Vendor Code	Name and Address of Vendor
V15.	Divinity Furnitures, No 435/01, Chariji Puram Main Road, Adyar Bus stop, Menambakkam, Chennai – 600022.	V16.	Blue Bull Chairs, Unit No21, Vitrus Bldg, Pune Highway Junction, Juhu, East.Mumbai – 23.
V17.	Jhalak Industries, E21, Anna Hazare Marg, Mumbai – 78.	V18.	Abhimanyu Enterprises, 753, Filmy Area, 3rd Road, Next block, Mumbai – 90.
V19.	Payal seaters, 23, Shamminagar, Dhobhighati, Mumbai – 22.	V20.	Kama Furniture, 36 Sastri Road, Naik Nagar, Vashi Road, Pune – 411022.
V21.	Selective Seating Systems, Opp. Joshi Hospital, Rampur Road, Savati, Pune – 411029.	V22.	Gadii Seating Systems, Opp Ram Tower Near Police Chowk, Shivaji nagar, Pune – 411045.
V23.	Kamble Furniturs, Avinasi RD, Dayaram Peth, Pune – 411089	V24.	Kartik Traders, Opp Rodrigo Company, Station Road, Pune – 411099.
V25.	Yakatulla Furnishings, near Telephone Exchange, Khadak Vasla, Pune – 411055		

The list of eligible vendors compiled by QSA is given below through the form shown as Fig. 7.4

This is the list of Items with vendors

Material Name	Material code	Vendor details
Material for hands	RM1	V1,V3,V12,V17,V25
Material for legs	RM2	V6,V11,V18,V19,V23
Material for Cover	RM3	V3,V5,V13,V18,V21
Seat plane	SP	V4,V8,V14,V16,V20
Fixture Sets	FX	V9,V13,V17,V21,V22
Rod	RD	V5,V7,V15,V19,V25

Fig. 7.4: Form for Material with Vendors

A sample form for sending the invitation for submitting quotation by QSA is shown below in Fig. 7.5. Similar quotations were sent for all the vendors. The responses for all these quotations are received by agent QRA which in turn forwarded only those vendors details to VRA who have sent sufficient information for the purpose of vendor rating. For Vendor rating VRA accepted weightages for various factors. For the proposed work cost, quality and transportation cost are considered.

Invitation for Quotation

To,
Vendor Name : Square Table Products

Address: Choukara palli, 49 ,11th S Main, 12th Cross , Bangalore – 560013.

Subject :Request for sending Quotation for the material listed Below
Reference No : 2011-2/34 DATED 12-08-2011

Sl.No.	Item Name	Item Code	Quantity	Due Date
1.	Material for Hand	RM1	700	
2.	Material for Cover	RM3	350	

You are requested to send your Quotation for the above Items at the earliest specifying following

Price [ ]   **Lead Time** [ ]

Quality [ ]

Transportation Modes and Cost [ ]

Note : The terms and conditions of the Purchasing are ttached as file

Signature

(Authorized Purchasing officer)

Fig. 7.5: Form for Inviting a Quotation

VRA performed vendor rating for each of the material to be purchased and prepared the list of selected vendors. For verification manual calculations were carried out for vendor rating. For material RM1 the calculations are shown in the Tables 7.6 and 7.7.

Table 7.6: List of Vendors for RM1

Vendor	Cost	Quality %	Distance KM.	Transportation Cost Rs./KM	Total Cost Rs.	Vendor Rating
V1	200	96	225	1	425	1
V25	210	97	300	1	510	2
V12	220	97	700	1	920	5
V17	200	96	560	1	760	3
V3	210	96	660	1	830	4
Vendor Rating: Weights Given C=40, Q=40, T=20						

Table 7.7: Vendor Rating for RM1

Vendor	Score for C	Score for Q	Score for T	Total score
V1	200/200 x 40 = 40	96/97 x 40 = 39.58	225/225 x 20 = 20	99.58
V25	200/210 x 40 = 38.1	97/97 x 40 = 40	225/300 x 20 = 15	93.10
V12	200/220 x 40 = 36.36	97/97 x 40 = 40	225/700 x 20 = 6.43	82.79
V17	200/200 x 40 = 40	96/97 x 40 = 39.58	225/560 x 20 = 8.04	87.62
V3	200/210 x 40 = 38.1	96/97 x 40 = 39.58	225/660 x 20 = 6.82	84.50
Vendor V1 is selected for RM1 (due to highest score)				

As a typical example processing performed by VRA for the material RM1 is shown in Fig. 7.6. The results generated by VRA is compared with manual calculations and found to be satisfactory. The Tables from 7.8 to 7.17 describe the calculations for other materials. The SVL generated by VRA which includes the list vendors selected for the placement of purchase orders is shown in Fig. 7.7.

Compiled Vendor Rating for Material RM1

Vendor	Address	Score for Quality	Score for Cost	Score for Distance	Score for Transportation	Total score
V1	A1	39.58	40	--	20	99.58*
V25	A2	40	38.1	-	15	93.10
V12	A3	40	36.36	--	6.43	82.79
V17	A4	39.58	40	------	8.04	87.62
V3	A3	39.58	38.1	--	6.82	84.50

Vendor V1 is selected since this vendor has performed better in terms of highest Total score*

Fig. 7.6: Sample Screen for Compiled Vendor Rating for a Material

Table 7.8: Vendors for RM2

Vendor	Cost ₹	Quality %	Distance KM.	Transportation Cost Rs./KM	Total Cost Rs.₹	Vendor Rating
V6	100	98	225	1	325	1
V11	120	95	700	1	810	5
V23	120	96	300	1	420	2
V18	100	97	660	1	760	3
V19	110	97	560	1	670	4
Vendor V6 is selected for RM2 (due to highest score)						

Table 7.9: Vendor Rating for RM2

Vendor	Score for C	Score for Q	Score for T	Total score
V6	100/100 x 40 = 40	98/98 x 40 = 40	225/225 x 20 = 20	100
V11	100/120 x 40 = 33.33	95/98 x 40 = 39.78	225/700 x 20 = 6.43	79.54
V23	100/120 x 40 = 33.33	96/98 x 40 = 39.18	225/300 x 20 = 15	87.51
V18	100/100 x 40 = 40	97/98 x 40 = 39.59	225/660 x 20 = 6.82	86.41
V19	100/110 x 40 = 36.36	97/98 x 40 = 39.59	225/560 x 20 = 8.04	83.99

Table 7.10: Vendors for RM3

Vendor	Cost ₹	Quality %	Distance KM.	Transportation Cost Rs./KM	Total Cost Rs.₹	Vendor Rating
V5	60	97	660	1	720	3
V21	65	96	300	1	365	2
V13	60	95	700	1	760	4
V18	70	98	560	1	630	5
V3	65	96	225	1	290	1
Vendor V3 is selected for RM3 (due to highest score)						

Table 7.11: Vendor Rating for RM3

Vendor	Score for C	Score for Q	Score for T	Total score
V5	60/60 x 40 = 40	97/98 x 40 = 39.59	225/660 x 20 = 6.82	86.41
V21	60/65 x 40 = 36.92	96/98 x 40 = 39.18	225/300 x 20 = 15	91.10
V13	60/60 x 40 = 40	95/98 x 40 = 38.78	225/700 x 20 = 6.43	85.21
V18	60/70 x 40 = 34.29	98/98 x 40 = 40	225/560 x 20 = 8.04	82.33
V3	60/65 x 40 = 36.92	96/98 x 40 = 39.18	225/225 x 20 = 20	96.10

Results and Discussions 185

Table 7.12: Vendors for Seating Plane (SP)

Vendor	Cost ₹	Quality %	Distance KM.	Transportation Cost Rs./KM	Total Cost Rs.₹	Vendor Rating
V4	150	96	225	1	375	1
V14	155	97	700	1	855	5
V16	150	98	560	1	710	4
V20	160	97	300	1	460	3
V8	150	95	225	1	375	2
Vendor V4 is selected for SP (due to highest score)						

Table 7.13: Vendor Rating for Seating Plane (SP)

Vendor	Score for C	Score for Q	Score for T	Total score
V4	150/150 x 40 = 40	96/98 x 40 = 39.18	225/225 x 20 = 20	99.18
V14	150/155 x 40 = 38.71	97/98 x 40 = 39.59	225/700 x 20 = 6.43	84.73
V16	150/150 x 40 = 40	98/98 x 40 = 40	225/560 x 20 = 8.04	88.04
V20	150/160 x 40 = 37.50	97/98 x 40 = 39.59	225/300 x 20 = 15	92.09
V8	150/155 x 40 = 38.71	95/98 x 40 = 38.78	225/225 x 20 = 20	97.49

Table 7.14: Vendors for Fixture (FX)

Vendor	Cost ₹	Quality %	Distance KM.	Transportation Cost Rs./KM	Total Cost Rs.₹	Vendor Rating
V21	50	95	300	1	350	3
V13	60	98	700	1	760	5
V17	55	97	560	1	615	4
V22	50	96	300	1	350	2
V9	55	97	225	1	280	1
Vendor V9 is selected for FX (due to highest score)						

Table 7.15: Vendor Rating for Fixture (FX)

Vendor	Score for C	Score for Q	Score for T	Total score
V21	50/50 x 40 = 40	95/98 x 40 = 38.78	225/300 x 20 = 15	93.78
V13	50/60 x 40 = 33.33	98/98 x 40 = 40	225/700 x 20 = 6.43	79.76
V17	50/55 x 40 = 36.36	97/98 x 40 = 39.59	225/560 x 20 = 8.04	83.99
V22	50/50 x 40 = 40	96/98 x 40 = 39.18	225/300 x 20 = 15	94.18
V9	50/55 x 40 = 36.36	97/98 x 40 = 39.59	225/225 x 20 = 20	95.95

Table 7.16: Vendors for Rod (RD)

Vendor	Cost ₹	Quality %	Distance KM.	Transportation Cost Rs./KM	Total Cost Rs.₹	Vendor Rating
V15	280	96	700	1	980	5
V19	280	97	560	1	840	3
V25	285	98	300	1	585	2
V7	285	95	225	1	510	1
V5	280	96	660	1	940	4
Vendor V7 is selected for RD (due to highest score)						

Table 7.17: Vendor Rating for Rod (RD)

Vendor	Score for C	Score for Q	Score for T	Total score
V15	280/280 x 40 = 40	96/98 x 40 = 39.18	225/700 x 20 = 6.43	85.61
V19	280/280 x 40 = 40	97/98 x 40 = 39.59	225/560 x 20 = 8.04	87.63
V25	280/285 x 40 = 39.29	98/98 x 40 = 40	225/300 x 20 = 15	94.29
V7	280/285 x 40 = 39.29	95/98 x 40 = 38.78	225/225 x 20 = 20	98.07
V5	280/280 x 40 = 40	96/98 x 40 = 39.18	225/660 x 20 = 6.82	86

Selected Vendors list (SVL)		
Material Name	Material code	Vendor details*
Material for Hand	RM1	V1
Material for Leg	RM2	V6
Material for Cover	RM3	V3
Seating Plane	SP	V4
Fixture Set	FX	V9
Rod	RD	V7

*The details will be obtained from the vendor database

Fig. 7.7: Sample Form for Material List with Selected Vendors

The agent POGA accepted the SVL as its major input and prepared purchase orders for all the selected vendors. As an example the purchase order for the material RM1 is shown in the Fig. 7.8. POGA dispatched all the purchase orders to the respective vendors.

Once, the materials are received from the vendors as per the purchase orders, the agent MRSUA ensured that the following activities have performed in time.

- Carrying out systematic inspection of the lots for their quality as well quantity verification.
- Sending the message to stores agent providing precise information about the result of inspection of lots and also the recommendations regarding the acceptance or rejection of lots.
- Updating of available quantity for a particular material, if the lot for that materials is accepted.

- Sending the information about the rejected lots to stores agent as well to respective vendors for further needful action.
- Updating the vendor database.

Purchase Order for Material

Name Company: A&U&K Furniture Manufacturing

Vendor Address: A&T ALL Works 5/8/1951 Fateh Multan Lane Maruti Ngr Cly Hyderabad – 500001.

Shipping Address: A&U&K Ltd Jevargi Road Industrial Area Shed No 22 Gulbarga

Item Code	Description	Quantity	Unit Price(Rs) ₹	Total(Rs) ₹
RM1	Material for Hand	700	200	140000
			Subtotal	140000
			Tax Rate: 12%	
			Tax: 16800	
			TOTAL:	156800
Mode of Transportation: Road				

Terms and Conditions: Refer Quotation

Fig. 7.8: Purchase Order for RM1

The discussions and calculation presented above shows that the agents developed in the proposed work have capabilities to carry out their functions automatically, leading to the effective sharing of information between various functional units of supply chain.

CHAPTER 8

CONCLUSIONS

8.1 SUMMARY

Supply Chain encompasses all those activities needed to design, make, deliver and use a product or service and therefore the pace of change and the uncertainty about how markets as well as global competition has made it increasingly important for companies to be aware of the supply chains they participate in and to understand the roles that they play. Since, the supply chain usually includes more than one company, the communication and information sharing between companies at the supplier-customer interface is critically important for overall supply chain performance. However, the major problem facing manufacturing organizations is how to provide efficient and cost-effective information sharing systems due to the complexity of supply chain networks coupled with the complexity of individual manufacturing systems within supply chains.

Concentrating on this important and critical issue, in the present work the significance of sharing information in supply chain activities has been highlighted and to meet the challenges of information sharing the basic concepts and characteristics of Multi Agent system approach has been evaluated and employed. By employing the concepts of agent based approach, a complete multi agent based frame work for information sharing in supply chain activities has been formulated and necessary agents have been identified and developed for performing major activities. For the purpose of effective information sharing agents are grouped into different modules, each group of agents in a module perform a major function of the enterprise.

The work mainly includes development of different Agents in the form of autonomous software, each capable of communicating with other Agents to share the information. Each pair of Agents communicates with each other by sending well defined messages and also responding suitably to any message received. The sharing of information between different units of supply chain has been accomplished through these agents The benefits of adopting agent technology in supply chain management are found to be several, especially it is possible to satisfy the major issues identified in the work regarding efficient flow of information in a supply chain. Therefore it can be concluded that the present work is expected to prove a significant step towards enhancing the efficiency and effectiveness of supply chain formed for manufacturing organizations.

8.2 SALIENT FEATURES

The following are the salient features of the work carried out:

- Through the literature survey it is shown that Information Technology plays a major role in the formation of the supply chain, also in enhancing the effectiveness and efficiency of supply chain activities.
- The Significance of sharing information between various units of Supply Chain for a Manufacturing organization has been highlighted and also the major issues related to sharing of information are identified.

- Various software tools have been evaluated for the purpose of information sharing and Agent Based Approach has been presented as one of the software development approach suitable for the same. It means to provide the necessary information technology support for the supply chains, the basic concepts and features of agent technology has been evaluated in order to determine its suitability for the management of supply chains.

- An overall framework or model has been identified for employing the basic concepts of Agent Based Approach leading to effective information sharing and enhancement of the capabilities of Supply Chain.

- The agents which are necessary to carryout collectively the various functions of MRP-I activities have been developed. Similarly individual sets of agents have been developed to carryout the functions related to Process planning and scheduling.

- For detailed Analysis the purchasing process has been exploited to divide it into interrelated steps which can be adopted for Agent Based Approach.

- The entire purchasing process has been automated through various Agents developed on VB dot net platform with required databases.

- A well defined database has been developed for storing information about Materials and Vendors which are expected to support for the automation of other activities also.

- The proposed work is focused on wide range of activities such as time phasing of material requirements, procurement of materials, manufacturing planning and shop floor control activities.

- For supplier selection an exclusive vendor database is maintained which is frequently updated on the basis of the past performance as well as any new additional information. Accordingly concerned agent assigns certain grades to each vendor. This leads to efficient vendor selection in comparison with conventional methods.

- The developed software interacts not only with internal system of the organisation such as stores data, vendor data etc. but also interacts with external system related to vendors, subcontractors etc.

8.3 LIMITATION OF THE PRESENT WORK

1. Real life cases/industrial problems are not employed in the present work as validation of work due to non-availability of necessary support and data. Further for validation a software package that supports Agent Based Approach is preferred.
2. Only purchasing aspect has been considered in depth for presenting the discussion on development of agents for flow of information.
3. Information flows related to inter enterprise activities are limited only with Vendors. More stress is given for intra enterprise activities.
4. The basic concept of multi agent system are used to develop the software and also forms for database as well as information flow on VB dot net platform. There are certain software platforms which exclusively support agent based approach and the work would have been more effective on such platforms.

8.4 SCOPE FOR FUTURE WORK

The present work can be extended to incorporate more effective features, in order to contribute more towards the enhancement of efficiency and effectiveness of supply chain activities through effective sharing of information, by concentrating on the limitations of the work listed above. In addition to that the following functional areas should be considered to carry out further research work on the development of necessary agents.

1. Communication with sub-contractors, service providers and retailers leading to inter organizational flow of information.
2. Transportation and distribution aspects.
3. Bringing necessary changes in the agents developed in the present work to deal with the dynamic shop floor conditions with realistic industrial practices.
4. By employing multi-objective optimization techniques to concentrate more on inventory management.
5. The current theory holds good for activities such as scheduling, manufacturing planning etc. and it is possible to extend the current work concentrating on these activities.

REFERENCES

[1] David J. Bloomberg, Stephen Lemay and Joe B. Hanna, "Logistics", prentice hall of India Pvt. Ltd., 2003.

[2] Sunil Chopra and Peter Meindl, "Supply Chain Management," 2001, Prentice Hall India.

[3] Marti' Nez-Olvera C. and Shunk D., "Comprehensive framework for the development of a supply chain strategy," 2006, International Journal of Production Research, Vol. 44, No. 21, pp 4511–4528.

[4] Lu T. P. Chang T. M and Yih Y., "Production control framework for supply chain management—an application in the elevator manufacturing industry," 2005, International Journal of Production Research, Vol. 43, No. 20, pp 4219–4233.

[5] Chan H. K. and. Chan F. T. S, "Effect of information sharing in supply chains with flexibility," 2006, International Journal of Production Research, pp1–20, preview article.

[6] Mario Verdicchio and Marco colombetti – "Commitments for Agent-Based Supply Chain Management ACM SIGecom Exchanges, Vol. 3, No. 1, 2002.

[7] Mark S. Fox, Mihai Barbuceanu, and Rune Teigen, "Agent-Oriented Supply-Chain Management," The International Journal of Flexible Manufacturing Systems, 12 (2000): 165–188.

[8] Uppin M S, Kamble Prashant G & Hebbal S S, "Information Technology for the processes of Manufacturing," Annals of DAAAM for 2005 & Proceedings – 16th DAAAM symposium on 'Intelligent Manufacturing & Automation: focus on young Researchers & Scientists' University of rueka, 19th – 22nd October 2005, Opatija, Croatia, pp 177 – 178.

[9] Danny C. K. Hok. F. Au and Edward newton, "Empirical research on supply chain management: a critical review and recommendations," 2002, Int. J. Prod. Res., vol. 40, no. 17, pp 4415±4430.

[10] Huang G. Q, Lau J. S. K., Mak K. L. and Liang L., "Distributed supply-chain project rescheduling: part I—impacts of information-sharing strategies," 2005, International Journal of Production Research, Vol. 43, No. 24, pp 5107–5129.

[11] Rasmus Friis Olsen, "Buyer-supplier relationships: alternative research approaches," 1997, European Journal of Purchasing & Supply Management, Vol. 3, No. 4, pp221–231.

[12] B.S. Sahay, "Supply Chain Management in twenty first century" – Macmillan India ltd, 2000.

[13] Weiming Shen and Douglas H. Norrie, "Agent-Based Systems for Intelligent Manufacturing: A State-of-the-Art Survey" URL: http://imsg.enme.ucalgary.ca/.

[14] Uppin M. S.and S. S. Hebbal, "Multi Agent System Model of Supply Chain for Information Sharing," International Journal of Contemporary Engineering Sciences, Vol. 3, 2010, No. 1, pp 1 – 16.

[15] Keah Choon Tan, "A framework of supply chain management literature," 2001, European Journal of Purchasing & Supply Management, Vol.7, pp. 39–48.

[16] Martin Christopher-"Logistics and Supply Chain Management Strategies for reducing cost and improving service," 2nd Edn. Pearson Education Delhi, 2003.

[17] Kun-Lin Hsieh, Yan-Kwang chen and Ching-Chen Shen, "Interval estimates of bullwhip effect based on bootstrapping technique," Journal of scientific & Industrial Research. Vol.66, October 2007, pp. 805–810.

[18] Richard Lancionia, Hope Jensen Schaub, Michael F. Smithc, "Internet impacts on supply chain management," 2003, Industrial Marketing Management, Vol.32, pp173–175.

[19] Richard Thomas, "Purchasing and technological change: Exploring the links between company Technology strategy and supplier relationships," 1994, European Journal of Purchasing and Supply Management Vol., No 3, pp.161–168.

[20] Truong T. H. and Azadivar F., "Optimal design methodologies for configuration of supply chains," 2005, International Journal of Production Research, Vol. 43, No. 11, pp 2217–2236.

[21] Simon Croom, Pietro Romano, Mihalis Giannakis, "Supply chain management: an analytical framework for critical literature review," 2000, European Journal of Purchasing & Supply Management, Vol.6, pp 67–83.

[22] Soo wook Kim and Ramnarasimhan, "Information system utilization in supply chain integration e. orts," 2002, Int. J. Prod. Res, vol. 40, no.18, pp 4585±4609.

[23] Park Y. B., "An integrated approach for production and distribution planning in supply chain management," 2005, International Journal of Production Research, Vol. 43, No. 6, pp 1205–1224.

[24] Larry C. Giuniperoa, Diane Denslowb, Reham Eltantawy, "Purchasing/supply chain management flexibility: Moving to an entrepreneurial skill set," 2005, Industrial Marketing Management, Vol. 34, pp 602 – 613.

[25] Lau J. S. K., Huang G. Q., Mak K. L. and Li L., "Distributed project scheduling with information sharing in supply chains: part II—theoretical analysis and computational study," 2005, International Journal of Production Research, Vol. 43, No. 23, pp 4899–4927.

[26] Angappa Gunasekaran and Bulent Kobu, "Performance measures and metrics in logistics and supply chain management: a review of recent literature (1995–2004) for research and applications," 2006, International Journal of Production Research, pp 1– 22, preview article.

[27] Ananda singgaram Jeeva, "Procurement dimensions in the Australian Mfg. sector: Flexibility issues in a supply chain perspective," Ph D thesis, School of Management, Curtin university of technology, Australia, March 2004.

[28] Tseng T.-L., Huang C.-C., Jiang F. and Ho J. C, "Applying a hybrid data mining approach to prediction problems: a case of preferred suppliers prediction," 2006, International Journal of Production Research, Vol. 44, pp 2935–2954.

[29] Chris Butterworth, "Supplier-driven partnerships," 1996, European Journal of Purchasing & Supply Management, Vol. 2, No. 4, pp. 169—172.

[30] Mehra S., "Current issues and emerging trends in supply chain management: an editorial perspective," 2005, International Journal of Production Research, Vol. 43, No. 16, pp 3299– 3302l.

[31] Ram Mudambi and Claus Peter Schriinder, "Progress towards buyer-supplier partnerships Evidence from small and medium-sized manufacturing firms," 1996, European Journal of Purchasing & Supply Management Vol. 2, No. 2/3, pp. 119–127.

[32] Y. Peng, T. Finin, Y. Labrou, B. Chu, J. Long, W. J. Tolone, A. Boughannam, "A Multi-Agent System for Enterprise Integration," www.citeseerx.ist.psu.edu/viewdoc/download?doi=10.1.1.39…

[33] B. Scholz-Reiter, H. Höhns, J. Kolditz, T. Hildebrandt, "Autonomous Supply Net Coordination," Proceedings of 38th CIRP Manufacturing Systems Seminar. Florianopolis, Brazil, 2005.pp 1-7.

[34] Yasuyuki Nishioka, "CAPPS: Collaborative Agents for Production Planning and Scheduling –A Challenge to Develop a new Software System Architecture for Manufacturing Management in Japan," 17th International Conference on Production Research, August 3–7, 2003, PDF 1–57.

[35] Hwarng H. B., Chong C. S. P., Xien N. and Burgess T. F., "Modelling a complex supply chain: understanding the effect of simplified assumptions," 2005, International Journal of Production Research, Vol. 43, No. 13, pp 2829–2872.

[36] Chen I. J. and Paulraj A., "Understanding supply chain management: critical research and a theoretical framework," 2004, Int. J. Prod. Res., vol. 42, no. 1, pp 131–163.

[37] Donald j Bowersox, David J. Closs, "Logistical management – The Integrated Supply Chain Process," Tata McGraw-Hill, 2005.

[38] Suwanruji P. and Enns S. T., "Evaluating the effects of capacity constraints and demand patterns on supply chain replenishment strategies," 2006, International Journal of Production Research, Vol. 44, No. 21, pp 4607–4629.

[39] Ste´phanie Hurtubise, Claude Olivier, Ali Gharbi, "Planning tools for managing the supply chain," 2004, International Journal of Computers & Industrial Engineering, Vol. 46, pp 763–779.

[40] Fynes B., de Bu´rca S. and Voss C., "Supply chain relationship quality, the competitive environment and performance," 2005, International Journal of Production Research, Vol. 43, No. 16, pp 3303–3320.

[41] Gwo-Hshiung Tzeng, Tzung-I Tang, Yu-Min Hung and Min-Lan Chang, "Multiple objective planning for a production and distribution model of the supply chain: Case of a bicycle manufacturer," Journal of scientific & Industrial Research. Vol.65, April 2006, pp. 309–320.

[42] Villegas F. A. and. Smith N. R, "Supply chain dynamics: analysis of inventory vs. order oscillations trade-off," 2006, International Journal of Production Research, Vol. 44, No.6, pp 1037–1054.

[43] W.J. Zhang, Q. Lib, "Information modelling for made-to-order virtual enterprise manufacturing systems," 1999, Computer-Aided Design, Vol.31, pp 611–619.

[44] Shih-Chia Changa, Rong-Huei Chenb, Ru-Jen Linc, Shiaw-Wen Tienb, Chwen Sheud, "Supplier involvement and manufacturing flexibility," 2006, Technovation, Vol.26, pp 1136–1146

[45] Vaidyanathan Jayaraman, "Production planning for closed-loop supply chains with product recovery and reuse: an analytical approach," 2006, International Journal of Production Research, Vol. 44, No. 5, pp 981–998

[46] Vedran Podobnik, Ana Petric and Gordan Jezic, "An Agent-Based Solution for Dynamic Supply Chain Management," Journal of Universal Computer Science, vol. 14, no. 7 (2008), pp 1080–1104.

[47] Damien J. Power, Amrik S. Sohal and Shams-Ur Rahman, "Critical success factors in agile supply chain Management An empirical study," 2001, International Journal of Physical Distribution & Logistics Management, Vol. 31 No. 4, pp247–265.

[48] Yogesh V. Joshi, "Information Visibility And Its Effect On Supply Chain Dynamics," Msc Thesis, Massachusetts Institute of Technology, June 2000.

[49] Tae-Young Kim, Sunjae Lee, Kwangsoo Kim, Cheol-Han Kim, "A modeling framework for agile and interoperable virtual enterprises," 2006, Computers in Industry, Vol.57, pp 204–217.

[50] P.K. Humphreys, M.K. Lai, D. Sculli, "An inter-organizational information system for supply chain management," 2001, Int. Journal of Production Economics, Vol.70, pp 245–255.

[51] Sebastia'n J. Garcı́a-Dastugue, Douglas M. Lambert, "Internet-enabled coordination in the supply chain," 2003, Industrial Marketing Management, Vol. 32, pp 251– 263.

[52] Mehmet Baruta, Wolfgang Faisstb, John J. Kanetc, "Measuring supply chain coupling: an information system perspective," 2002, European Journal of Purchasing & Supply Management, Vol. 8, pp 161–171.

[53] Christopher P Holland, "Cooperative supply chain management: the impact of inter organizational information systems," 1995, Journal of Strategic Information Systems, Vol. 4, No.2, pp 117–133.

[54] Chen H.-Y. S., Lin C.-W. R. and Yih Y., "Production-distribution network design of a global supply chain alliance from the key player's perspective," 2006, International Journal of Production Research, pp 1–21, preview article.

[55] Dominguez H. and Lashkari R. S., "Model for integrating the supply chain of an appliance company: a value of information approach," 2004, Int. J. Prod. Res., vol. 42, no. 11, pp 2113–2140.

[56] Chan F. T. S., "Interactive selection model for supplier selection process: an analytical hierarchy process approach," 2003, Int. J. Prod. Res., vol. 41, no. 15, pp 3549–579.

[57] K. Lau J. S., Huang G. Q., Mak K. L. and Liang L.," Distributed project scheduling with information sharing in supply chains: part I—an agent-based negotiation model," 2005, International Journal of Production Research, Vol. 43, No. 22, pp 4813–4838.

[58] Qing zhang, "Essentials for Information coordination in supply chain systems," journal of asian social science, vol.4, no. 10, October 2008, www.ccsenet.org/journal.html.

[59] Lyons A. C., Coronado Mondragon A. E., Breman A., Kehoe D. F. and Coleman J., "Prototyping an information system's requirements architecture for customer-driven, supply-chain operations," 2005, International Journal of Production Research, Vol. 43, No. 20, pp 4289–4319.

[60] Mo J.P.T., Zhou M., "Tools and methods for managing intangible assets of virtual enterprise," 2003, International journal of Computers in Industry, Elsevier Science Ltd. Vol. 51, pp 197– 210.

[61] Martin Grieger, "Electronic marketplaces: A literature review and a call for supply chain management research," 2003, European Journal of Operational Research, Vol. 144, pp 280–294.

[62] Chrwan-Jyh Ho, "Measuring system performance of an ERP-based supply chain," 2006, International Journal of Production Research, pp 1–23, preview article.

[63] Craighead C. W., Patterson J. W., Roth P. L. and Segars A. H., Enabling the benefits of Supply Chain Management Systems: an empirical study of Electronic Data Interchange (EDI) in manufacturing," 2006, International Journal of Production Research, Vol. 44, No. 1, pp 135–157.

[64] Richard A. Lancioni, Michael F. Smith, Hope Jensen Schau, "Strategic Internet application trends in supply chain management," 2003, Industrial Marketing Management, Vol.32, pp 211 – 217.

[65] Monczka, Trent, Handfield – "Purchasing and Supply Chain Management" 2nd Edn. Thomson Asia pte Ltd, Singapore 2002.

[66] Rahman Zillur, "Internet-based supply chain management: using the Internet to Revolutionize your business," 2003, International Journal of Information Management, Vol.23, pp 493–505.

[67] Monthatipkul C. and Kawtummachair., "Algorithm for constructing a delivery sequencing/inventory-allocation plan for supply chain control in the operational planning level," 2006, International Journal of Production Research, 1–21, preview article.

[68] Noorul Haq A. and Kannan G., "Design of an integrated supplier selection and multi echelon distribution inventory model in a built-to-order supply chain environment," 2006, International Journal of Production Research, Vol. 44, No. 10, pp 1963–1985.

[69] Ricardo, R. Fernandez, "Total Quality in purchasing & supplier management," Productivity press (India) Pvt. Ltd., 1996.

[70] Green JR K. W. and Inman R. A., "Using a just-in-time selling strategy to strengthen supply chain linkages," 2005, International Journal of Production Research, Vol. 43, No.16, pp 3437–3453.

[71] Tanvi Kotharia, Clark Hub and Wesley S. Roehl, "Adopting e-Procurement technology in a chain hotel: An exploratory case study," 2006, Hospitality Management, Elsevier Ltd.

[72] Teck-Yong Eng., "The role of e-marketplaces in supply chain management," 2004, Industrial Marketing Management, Vol.33, pp 97– 105.

[73] Terry Anthony Byrda, Nancy W. Davidson, "Examining possible antecedents of IT impact on the supply chain and its effect on firm performance," 2003, Information & Management, Vol.41, pp 243–255.

[74] William D. Presutti Jr., "Supply management and e-procurement: creating value added in the supply chain," 2003, Industrial Marketing Management, Vol.32, pp 219– 226.

[75] Lisa M. Huntera, Chickery J. Kasouf, Kevin G. Celuch, Kathryn A. Curry, "A classification of business-to-business buying decisions: Risk importance and probability as a framework for e-business benefits," 2004, Industrial Marketing Management, Vol. 33, pp 145– 154.

[76] Robert B. Handfielda, Christian Bechtelb, "The role of trust and relationship structure in improving supply chain responsiveness," 2002, Industrial Marketing Management, Vol.31, pp 367– 382.

[77] Daniel J. Flint, "Strategic marketing in global supply chains: Four challenges," 2004, Industrial Marketing Management, Vol. 33, pp 45– 50.

[78] Louis Raymond and Samir Blili, "Adopting EDI in a network enterprise: the case of subcontracting SMEs," 1997, European Journal of Purchasing & Supply Management, Vol. 3, No. 3, pp. 165–175.

[79] Jens Laage-Heliman and Lars-Erik Gadde, "Information technology and the efficiency of materials supply: The implementation of EDI in the Swedish construction industry," 1996, European Journal of Purchasing & Supply Management, Vol. 2, No. 4, pp. 221–228.

[80] Caridi M., Cigoloni R. and De marco D., "Improving supply-chain collaboration by linking intelligent agents to CPFR," 2005, International Journal of Production Research, Vol. 43, No. 20, 4191–4218.

[81] Massimo Paolucci Roberto Sacile, "Agent based Manufacturing and control systems – New Agile Manufacturing solutions for achieving peak performance," CBC Press, 2005.

[82] Weiming Shen, Rob Kremer, Mihaela Uieru and Douglas Norrie, "A collaborative agent-based infrastructure for Internet-enabled collaborative enterprises," 2003, Int. J. Prod. Res., vol. 41, no. 8, pp 1621–1638.

[83] Charles M. Macal, Michael J. North, "TUTORIAL ON AGENT-BASED MODELING AND SIMULATION," Proceedings of the 2005 Winter Simulation Conference, pp 1–14.

[84] David Zhengwen Zhang, Anthony Ikechukwu Anosike, Ming Kim Lim, Oluwaremilekun Mowanuola Akanle, "An agent-based approach for e manufacturing and supply chain integration," 2006, International Journal of Computers & Industrial Engineering.

[85] Iman Badr, "An Agent-Based Scheduling Framework for Flexible Manufacturing systems," International Journal of Computer, Information, and Systems Science, and Engineering 2;2 www.waset.org Spring 2008, pp 123 – 129.

[86] Mihai Barbucean, rune Teigen and Mark s. Fox, "Agent Based design and Simulation of Supply chain systems," Proceedings of WETICE'97, IEEE Computer society press, pp 36–42, 1997.

[87] Ask Just Jensen, Kasper Hallenborg, Yves Demazeau, "Reactive agent mechanisms for scheduling manufacturing processes," AT2AI-6 Working Notes, from Agent Theory to Implementation, 6th Int. Workshop, May 13, 2008, AAMAS 2008, Estoril, Portugal, EU.

[88] Uppin M.S and S.S. Hebbal, "Agent based framework for supply chain management," International conference on Advanced Manufacturing and Automation (INCAMA2009), Kalasamangalam University, TN, India, 26–28 march 2009.

[89] Narayan Rangaraj, "Modelling in e-Business and Supply Chain Management," Invited paper for the Workshop on Frontiers of E-business, organized by GM-India Science Labs, Bangalore, December 2005.

[90] Agostino Villa, "The Norman Dudley Paper – Emerging trends in large-scale supply chain management," 2002, Int. J. Prod. Res., vol. 40, no.15, pp 3487±3498.

[91] Allwood J. M. and Lee J.-H., "The design of an agent for modelling supply chain network dynamics," 2005, International Journal of Production Research, Vol. 43, No.22, pp 4875–4898.

[92] Weiming Shen, Douglas H. Norrie, J-P. Barthes, "Multi-Agent Systems for Concurrent Intelligent Design and Manufacturing," [eBook], http://books.google.co.in/books

[93] Oluwaremilekun Mowanuola Akanle, "An agent-based approach for e-manufacturing and supply chain integration," 2006, International Journal of Computers & Industrial Engineering.

[94] Bob Roberts, Mike Mackay, "IT supporting supplier relationships: The role of electronic commerce," 1998, European Journal of Purchasing & Supply Management, Vol.4, pp 175–184.

[95] George Q. Huang, Jason S. K. Lau and K. L. Mak, "The impacts of sharing production information on supply chain dynamics: a review of the literature," 2003, Int. J. Prod. Res., vol. 41, no. 7, pp 1483–1517.

[96] José Barata, L.M. Camarinha-Matos, Raymond Boissier, Paulo Leitão, Francisco Restivo & Mohammed Raddadi, "Integrated and distributed manufacturing, a Multiagent perspective," http://www.ipb.pt/~pleitao/papers/wesic2001.pdf.

[97] Rolón, M. et al. (2009), "Agent Based Modelling and Simulation of Intelligent Distributed Scheduling Systems," Proceedings of the 19th European Symposium on Computer Aided Process Engineering – ESCAPE19, Elsevier B.V.

[98] Bastin Tony Roy Savarimuthu, Maryam Purvis, Martin Purvis, "Agent Based Web Service Composition In The Context Of A Supply-Chain Based Workflow," The Information Science Discussion Paper Series Number 2006/04 February 2006 ISSN 1172–6024, Department of Information Science Uni of Otago, NZL.

[99] Cakravastla and K. Takahashi A, "Integrated model for supplier selection and negotiation in a make-to-order environment," 2004, Int. J. Prod. Res., vol.42, no. 21, pp 4457–4474.

[100] Venkateswaran J. and. Son Y. J, "Impact of modelling approximations in supply chain analysis – an experimental study," 2004, Int. J. Prod. Res., vol. 42, no. 15.

[101] Campbell J. and Sankaran J., "An inductive framework for enhancing supply chain integration," 2005, International Journal of Production Research, Vol. 43, No. 16, pp 3321– 3351.

[102] Yee S.-T., "Impact analysis of customized demand information sharing on supply chain performance," 2005, International Journal of Production Research, Vol. 43, No. 16, pp 3353– 3373.

[103] Haitham Al-zu'bi, "Applying Electronic Supply Chain Management Using Multi-Agent System: A Managerial Perspective," International Arab Journal of e-106 Technology, Vol. 1, No. 3, January 2010, pp 106 – 113.

[104] Dr. Bo Zhao, Prof. Dr. Yushun Fan, "Multi-agent Based Integration of Scheduling Algorithms," Proceedings of the IASTED International Conference-Intelligent Systems and Control, 2001.11, pp 55–59.

[105] G. Balakayeva, and A. Aktymbayeva, "Multiagent Systems Simulation," World Academy of Science, Engineering and Technology 28, 2007, pp 1–3. http://www.waset.org/journals/waset/v28/v28-1.pdf.

[106] Jos´e M Vidal, Paul Buhler, Hrishikesh Goradia, "The Past and Future of Multiagent Systems," AAMAS Workshop on Teaching Multi-Agent Systems, 2004, pp 1 – 8.

[107] Katia P. Sycara, "Multiagent Systems," SUMMER 1998, American Association for Artificial Intelligence, pp 79–83, www.perada.eu/documents/articlesperspectives/multi-agent-systems. PDF.

[108] Roberto A. Flores-Mendez, "Towards a Standardization of Multi-Agent System Frameworks," 1999, http://www.acm.org/crossroads/xrds5-4/multiagent.html.

[109] Olaf Bochmann, "Probabilistic Approaches in Multi-Agent Systems for Manufacturing Coordination and Control," Katholieke university, Arenbergkasteel, B-3001 Heverlee (Leuven), Belgium, June 2005.

[110] Vachon S. and Klassen R. D., "Supply chain management and environmental technologies: The role of integration," 2006, International Journal of Production Research, pp, 1–23, preview article

[111] Srinivasan G, "Quantitative Models in Operations and Supply chain management," PHI learning private limited, 2010.

[112] Heiko Wolters, Frank Schuller, "Explaining supplier-buyer partnerships: a dynamic game theory approach," 1997, European Journal of Purchasing & Supply Management, Vol. 3, No. 3, pp. 155–164.

[113] Srinivas Talluri and Joseph Sarkis, "A model for performance monitoring of suppliers," 2002, Int. J. Prod. Res., vol. 40, no.16, pp 4257±4269.

[114] Jack C. Hayya, Jeon G. Kim, Stephen M. Disney, Terry P. Harrison and Dean hatfield, "Estimation in supply chain inventory management," 2006, International Journal of Production Research, Vol. 44, No. 7, pp 1313–1330.

[115] Dotoli M., Fanti M. P., Meloni C. and Zhou M. C., "A multi-level approach for network design of integrated supply chains," 2005, International Journal of Production Research, Vol. 43, No. 20, pp 4267–4287.

[116] Zsidisin G. A., Melyk S. A. and Ragatz G. L., "An institutional theory perspective of business continuity planning for purchasing and supply management," 2005, International Journal of Production Research, Vol. 43, No. 16, pp 3401–3420.

[117] Uppin M. S, Hebbal S. S, "An overview of Agent Based model for Supply chain management," International journal of Emerging Technologies and Applications in Engineering Technology and Science, Vol. 3: issue 1, jan' 10 – jun' 10, 2010. pp 313 – 317.

[118] H K Gules and T F Burgess, "Manufacturing technology and the supply chain Linking buyer-supplier relationships and advanced manufacturing technology," 1996, European Journal of Purchasing & Supply Management Vol. 2, No 1, pp. 31–38.

[119] O'Donnel T., Maguire L., Mcivor R. and Humphrey P., "Minimizing the bullwhip effect in a supply chain using genetic algorithms," 2006, International Journal of Production Research, Vol. 44, No. 8, pp 1523–1543.

[120] Eleni Mangina, "Intelligent Agent-Based Monitoring Platform for Applications in Engineering," International Journal of Computer Science & Applications © 2005 Technomathematics Research Foundation, Vol. 2, No. 1, pp. 38 – 48.

[121] Erhan kutanoglu and s David Wu, "Incentive compatible, collaborative production scheduling with simple communication among distributed agents," International Journal of Production Research, Vol. 44, No. 3, 1 February 2006, pp 421–446.

[122] Kamble P G, Sambrani A and Hebbal S S – "Agent based approach for supply chain management," Proceedings of the national conference on Modern trends in management, Belgaum, 2004

[123] Rasoul Karimi, Caro Lucas, Behzad Moshiri, "New Multi Attributes Procurement Auction for Agent-Based Supply Chain Formation," IJCSNS International Journal of Computer Science and Network Security, VOL.7 No.4, April 2007.

[124] Sanjeev Reddy, K.H., Dr. Prahlada Rao, K. "Just-In–Time Manufacturing System with Lot Size and Set up Time Optimization," International Journal of Computer Mathematical Sciences and Application (IJCMSA). vol. 2 (2), (2008), pp 89–96.

[125] Sanjeev Reddy, K.H., Dr. Prahlada Rao, K, "Total Productive Maintenance (TPM): An Operational Efficiency Tool," International Journal of Business Policy and Economics. vol. 3 (2), July 2011.

[126] Uppin M.S., Kamble Prashant G. and S.S. Hebbal, "Realisation of virtual enterprise through the formation of supply chain," Proceedings of International conference on Advanced materials, design & manufacturing systems, Bannari Amman Institute of Technology, Satyamangalam, TN, India, 12-14 Dec 2005.

M S Uppin

Author's Publications

PUBLICATIONS IN INTERNATIONAL JOURNALS

1. Uppin M. S, Hebbal S. S, "An overview of Agent Based model for Supply chain management," International journal of Emerging Technologies and Applications in Engineering Technology and Science, Vol. 3: issue 1, Jan '10 – Jun '10, 2010. pp 313–317.

2. Uppin M. S. and S. S. Hebbal, "Multi Agent System Model of Supply Chain for Information Sharing," International Journal of Contemporary Engineering Sciences, Vol. 3, 2010, No. 1, pp 1–16.

3. Uppin M. S. and S. S. Hebbal, "Agent Model of Supply Chain for Process Planning," International journal of Engineering Research and Industrial Applications (IJERIA), Vol. 4, No. I (February 2011), pp 101–112.

4. Uppin M. S., Shivkumar Gosul and S. S. Hebbal, "Multi Agent Model for Automation of Purchasing Process," International Journal of Emerging Technologies and Applications in Engineering Technology and Science, Jul-Dec, 2011, Vol. 4, No. 2, pp 240–245.

5. Uppin M S and Hebbal S S, "Multi Agent Framework for Purchasing Process," International Journal of Applied Engineering Research, Vol. 6, No. 20(2011), pp. 2431–2440.

PUBLICATION IN INTERNATIONAL CONFERENCES

1. Uppin M S, Kamble Prashant G & Hebbal S S, "Information Technology for the processes of Manufacturing," Annals of DAAAM for 2005 & Proceedings – 16th DAAAM symposium on 'Intelligent Manufacturing & Automation: focus on young Researchers & Scientists' University of rueka, 19th – 22nd October 2005, Opatija, Croatia, pp 177–178.

2. Uppin M.S., Kamble Prashant G. and S.S. Hebbal, "Realisation of virtual enterprise through the formation of supply chain," Proceedings of International conference on Advanced materials, design & manufacturing systems, Bannari Amman Institute of Technology, Satyamangalam, TN, India, 12–14 Dec 2005.

3. Uppin M.S and S.S. Hebbal, "Agent based framework for supply chain management," International conference on advanced manufacturing and automation (INCAMA2009), Kalasamangalam, (T.N.) India, 26–28 march 2009.

www.ingramcontent.com/pod-product-compliance
Lightning Source LLC
Chambersburg PA
CBHW030921180526
45163CB00002B/419